The Art and Craft of Papermaking

The Art and Craft of Papermaking

Sophie Dawson

RUNNING PRESS
PHILADELPHIA, PENNSYLVANIA

A QUARTO BOOK

9 8 7 6 5 4 3 2
Digit on the right indicates the number of the printing.

ISBN 1-56138-158-6

Library of Congress Cataloging-in-Publication Number 92-50186

This publication was designed and
produced by Quarto Publishing plc
6 Blundell Street
London N7 9BH
England

Designer: Graham Davis
Art editor: Penny Cobb
Project editor: Mary Senechall
Senior editor: Sally MacEachern
Photographers: Jon Wyand, Paul Forrester
Picture researcher: Liz Eddison
Art director: Moira Clinch
Publishing director: Janet Slingsby
Typeset by En-to-En, Kent
Manufactured in Hong Kong
by Regent Publishing Services, Ltd
Printed in Hong Kong by Leefung Asco Ltd

This book may be ordered from the publisher.
Please add $2.50 postage and handling. *But try your bookstore first!*

Running Press Book Publishers
125 South Twenty-second Street
Philadelphia, Pennsylvania 19103

Contents

1

6 INTRODUCTION

2

16 GETTING STARTED
18 Papermaking: East and West
20 Mold and deckle
23 Sources of fiber
25 Preparing the fiber
30 Beating and pulp preparation
36 Vats
38 Sheetforming and couching
44 Pressing and parting
48 Drying the paper
51 Sizing and finishing

3

54 PAPERMAKING VARIATIONS
56 Laminating
59 Embedding
62 Flowers and foliage
65 Decorative effects
69 New outlines: making shaped paper
72 Embossing
74 Collage and combining techniques
78 Paper's precursors
82 Watermarks: a hidden image
85 Pulp painting
88 Pulp spraying
91 Vacuum-forming
95 Gallery

WARNING

A word of warning about dyes, pigments and other additives: Many chemical additives can be quite toxic and are poisonous if ingested or inhaled. Some can cause skin irritations. Precautions are, therefore, urged in dealing with any chemical compound. Good ventilation and the use of a mask are recommended when using powdered pigments; splash goggles and rubber gloves should also be used when mixing stock solutions. The use of a skin guard (barrier cream) may also be appropriate.

4

98 SCULPTURAL TECHNIQUES
100 Cast paper
110 Gallery

5

112 CONTEMPORARY ADAPTATIONS
114 Paper and books
122 Paper and textiles
126 Paper and light
129 Paper and nature
134 Paper and prints
137 Gallery

140 Glossary
142 Index
144 Papermaking suppliers
144 Acknowledgments

CHAPTER

1

Introduction

Julia Manheim: Guardian Angel. Front pages of The Guardian *newspaper and wire, (right).*

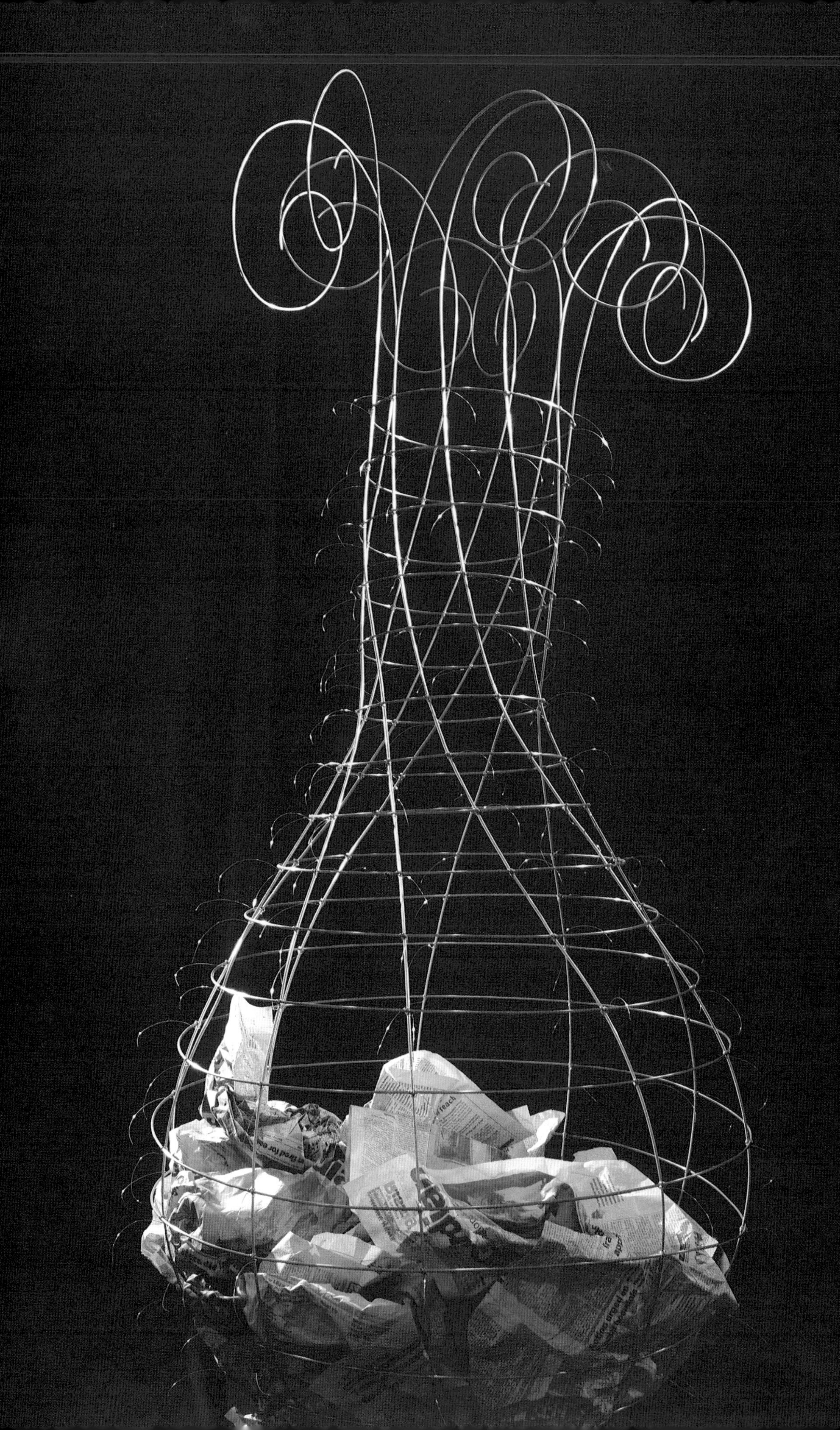

Paper can assume many forms, and the methods of papermaking, along with its historic uses in various cultures, have been remarkably different. Paper has been respected as a sacred material in the East and simultaneously employed there for a variety of utilitarian purposes, from making firecrackers to floor coverings. In the West, a more generally pragmatic approach to paper was adopted. This was largely due to the timing of its introduction to Europe, a millennium after its invention in China. It was predominantly used for writing, printing, and wrapping, in the form of newspapers, book pages, illustration papers, bank notes, artists' papers, and wallpapers.

With the advent of industrialization, paper lost the inherent features of a finely hand-crafted material. Paradoxically, the technology that made its mass production possible – the invention of chemical bleaching and acid sizing and the use of ground wood pulp – became a severe threat to the permanence of paper. Its role as a unique surface for writing, drawing, painting, and printing has been further undermined by the advance of electronic communications media: paper has become one of the major disposables of contemporary society.

A reaction against mass production and greater ecological awareness has, however, led to a renewed interest in both Asian and Western methods of hand papermaking. On one hand, there has been a gradual revitalization of the craft and a reaffirmation of traditional methods. On the other, there has been an unprecedented level of direct experimentation by artists from a variety of disciplines, who have extended paper's esthetic role far beyond the limits of its accustomed use.

ANCIENT PRECURSORS

The search for a material capable of transmitting and preserving written information has led to a great variety of solutions. Some were temporary measures; others were meant to guarantee permanence. Stone, metal, wood, waxed tablets, and clay tablets with their own

Egyptian sarcophagus of Rameses III (below right). Chinese "spirit money" is burned at funerals to placate ancestral deities (right). Josephine Tabbert: Lager aus Archäologischer (detail). Handmade paper (left).

clay envelopes were used in many ancient civilizations. They were gradually replaced by flexible alternatives that were cheaper to prepare and less cumbersome to transport.

Papyrus, made from a plant of that name, was the chief writing material in Egypt and the Greco-Roman world for almost 4,000 years. It was used for literary and administrative documents, receipts, petitions, and private and official letters. Our word "paper" derives from the Greek "papyros," in turn probably borrowed from Egypt.

Parchment, made from the treated skins of goats and sheep, was developed by the Persians as an alternative to papyrus. It was more durable and flexible than papyrus and was used for works of literary or religious importance.

Palm leaves and birch bark were the most common forms of writing material in India and Southeast Asia, where bamboo and wood were also used. Palm leaves were the easiest to process into a suitable form for writing and were bound together into books by a cord threaded through each leaf.

THE INVENTION OF PAPER

Paper was invented by the Chinese during the Han dynasty (202 B.C.-A.D. 220). Its "discovery" in A.D. 105 is credited to an Imperial court official named T'sai Lun. Prior to the development of paper, writing had generally been done on bamboo slips or pieces of silk. Bamboo was heavy and awkward to transport and silk was expensive; so paper, being lighter and cheaper to produce, was a welcome replacement. Paper was initially of poor quality, made from tree bark, hemp remnants, rags, and fishing nets. But as papermaking methods were refined, it became an accepted and favored writing surface.

The value placed on learning and literature in Chinese culture created an enormous demand for writing materials. Paper served the administrative needs of China's developing bureaucracy and later became the means of economic management with the issue of paper currency.

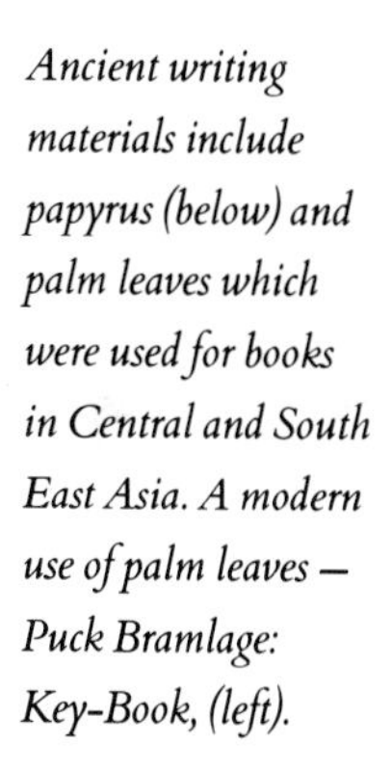

Ancient writing materials include papyrus (below) and palm leaves which were used for books in Central and South East Asia. A modern use of palm leaves – Puck Bramlage: Key-Book, (left).

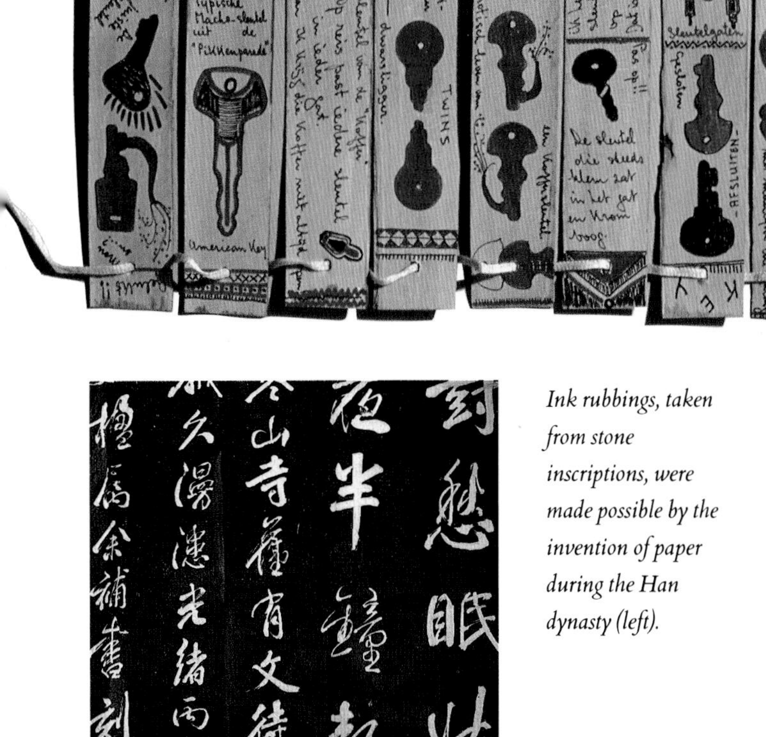

Ink rubbings, taken from stone inscriptions, were made possible by the invention of paper during the Han dynasty (left).

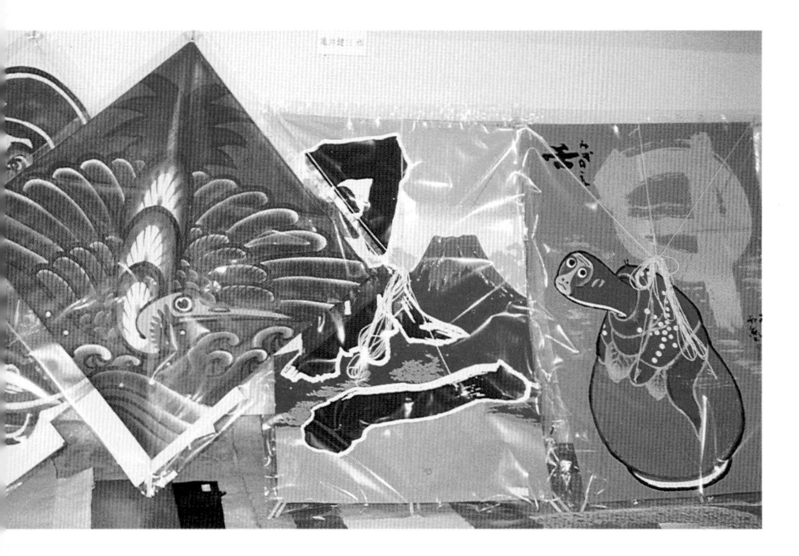

Kite paper is remarkable for its strength and flexibility, demonstrated in the early kites used for martial purposes in China and Korea. Festival, ceremonial, and fighting kites are a colorful element of holiday celebrations.

Paper also played a prominent part in Chinese religious practices. Ornamental papers fashioned in the form of symbolic objects were burned at Chinese funerals, and paper representations of Buddhist and Taoist gods were pasted onto walls to protect the home from misfortune.

THE DUAL ROLE OF PAPER

The art of papermaking was a closely guarded secret for almost 500 years. It was introduced to Japan through Korea at about the same time as Buddhism, early in the seventh century. As in China, paper was used for the copying of sutra, or sacred Buddhist texts. In A.D. 770 the Empress Shotoku commissioned one million printed paper prayers (*dharani*), each enshrined in a miniature wooden pagoda, to be distributed to Buddhist temples.

The use of paper as a sacred material was readily incorporated into many Japanese Shinto ceremonies. Paper was often used as a symbol of purity and had a spiritual significance that was an inseparable aspect of daily life. Paper ornaments and small strips of paper are still used as talismans and are hung up to mark sacred places.

Paper also had many esthetic and functional uses. It began to be decorated, dyed, patterned, and to appear in more and more varieties, while the number of utilitarian products made from it also increased. In Korea, for example, paper was oiled to make floor coverings, lacquered to make chests, and used as sails for boats.

The strength, translucency, and flexibility characteristic of *washi* (Japanese handmade paper) allowed an astonishing number of secular uses. Paper was not only used for writing, calligraphy, books, and printing with colored woodblocks; it was also made into fans, umbrellas, bags, banners, masks, kites, lanterns, tarpaulins, clothing, and sliding screens (*fusama*) and windows (*shoji*).

THE MIGRATION OF PAPER

During an attack on the city of Samarkand in central Asia in A.D. 751, the Arabs are reported to have taken many Chinese prisoners, some of whom

Tobacco pouches (right) made from thick, oiled paper to resemble leather. To emboss the striped pattern on the left-hand pouch, the paper was probably dampened and then pressed into a carved woodblock.

Woodblock prints depicting scenes from the life of the Buddha were frequently used as frontispieces for Chinese Buddhist works and clearly reveal their Indian origin.

Parchment, made from the prepared skin of animals, and finer quality vellum, provided a fine, smooth writing material and superseded papyrus and wax tablets (right).

are said to have been papermakers. From Samarkand the art of papermaking spread throughout the Islamic world. Important manufacturing centers were established in Baghdad, Damascus, and Cairo; and by the ninth century, paper was preferred to papyrus and parchment as a writing material.

Paper was available in Europe as an expensive import from the Arabs several hundred years before its manufacture in

INCIPIT TERTIUSDECIMUS DE PEREGRINIS
ACTENV
odoribus habe
silue Frantor
singula. uiuit
ea omnia mi
cunctis unum
cere: ita reper
guenta
mu de ungu
quomodo
ta nob & d
tione coru
Ngu
prim
non traditum
poribus non er
thure supplicabatur. cedri tantum & citri suorum fruti
sacris fumo conuolutum in nidore uerius odorem nouer

local mills. At this time Europeans were using parchment and its finer equivalent, vellum (made from calfskin, lambskin, or kidskin); but as elsewhere, once paper became a local product, it displaced these more expensive materials. The European paper industry started in Spain, where recent research shows there were important manufacturing centers in Cordoba, Seville, and Xativa. By 1276 the craft of papermaking had reached Fabriano, Italy. It spread slowly through Europe during the 14th and 15th centuries, and in 1690, the first paper mill was established in the North American colonies.

THE CHANGING FACE OF PAPER

Although the earliest uses of paper in the West were primarily religious, (Christian manuscripts, votive images of the life of Christ, indulgences for pilgrims), the increase in literacy that accompanied the Renaissance and the decline in power of the Church put a different emphasis on the role of paper.

The development of new printing processes, notably the invention of a movable typeface by Johann Gutenberg (*c* 1446), advanced paper's importance in the publishing of books and periodicals. New illustrative methods, such as woodcut, engraving, etching, and mezzotint, extended the esthetic use of paper in printed images.

The imperatives of an expanding commerce created a huge demand for quality paper to be used for certificates, diplomas, deeds, bank notes, and receipts as well as book publishing, while cheaper grades were needed for wrapping and packaging. The versatility of paper was demonstrated in 17th- and 18th-century England as an increasing number of mills began to manufacture varieties of card and board for papier-mâché, japanned ware, and playing cards, which were made from sheets of paper pasted together. Rising demand, the development of new types of paper, and a growing realization of the profitability

An antique laid mold and deckle from the oldest existing (13th century) paper mill in Amalfi, Italy, (below).

Japanese folded paper fans (above) were probably first made during the 12th century. The flat uchiwa *fan also became an indispensable popular accessory, and both are still in widespread use more than 800 years later.*

Engraving (1700), depicting the interior of a paper mill showing a set of water-powered wooden stampers used for pounding and refining the rags, the vat, couching, and press.

of papermaking led to the emergence of a major industry.

CHANGES IN MANUFACTURE

In the course of its migration to Europe, the materials and production methods for making paper changed. These changes kept output in line with demand and satisfied the requirements of its end use. Paper was initially made by pouring pulp onto a woven cloth that was stretched across a wooden frame. The newly formed sheet was removed from the cloth only after it had dried, and many such molds were needed to produce a useful quantity of paper.

Production was increased by the development of a mold with a removable bamboo or reed-grass cover, which allowed the paper to be transferred and dried off the mold. The Japanese used this to develop their own unique method of making paper called *nagashi-zuki* (see p. 18).

Paper made from kozo and gampi, the inner bark fibers used for papermaking in Japan, was well suited to calligraphic writing with a soft brush and compared quite favorably with hemp-based paper from China and Korea.

Old cotton and linen rags, and hemp fibers from ropes and sailcloth, provided the raw materials for Western papermaking; and, following improved Middle Eastern methods for pounding fibers, a water-driven stamper was devised to beat and defiber the cloth. Writing with a European quill pen and oxgall inks required a different surface from that needed in the East. Therefore new methods of sizing and finishing the sheets were introduced to produce a smoother, harder, and more opaque paper.

Further changes were also made to the construction of papermaking molds. It is likely that early Spanish and Italian papers were made on molds which closely resembled those used in Eastern papermaking. In Europe, however, there was no suitable natural substitute for the Eastern materials, such as bamboo or thin strips of dried grass. The art of drawing metal into a slender thread or rod by hand is thought to have been practiced in Italy at the beginning of the 12th century; and metal wires, possibly made of copper, soon replaced earlier materials. The configuration of closely spaced parallel "laid" lines held together by vertical stitching threads, or "chain" lines, remained essentially unchanged until the introduction of the European "wove" mold in the mid-18th century. Both mold covers differed considerably from their Eastern counterpart in that they were permanently attached to the mold frame.

During the 17th century, Holland took the lead in paper production, and the quality of Dutch paper was highly regarded. The Hollander beater, designed toward the end of the century and named after its country of origin, represented a significant advance in mechanical methods of refining rags for papermaking.

While these improvements made paper production more economical and led to its quantitative growth, the supply of rags could by no means equal demand. The shortage of raw materials generated an extensive search for alternative fibers and precipitated the change to wood pulp as the basic raw material for Western papermaking in the 19th century.

In the 17th century, a new mechanical device for beating pulp was invented in Holland. It is now widely known as the "Hollander beater."

MECHANIZATION

Until the end of the 18th century, all paper was made by hand. But with the invention by Nicholas Louis Robert in 1798 of a papermaking machine that could in principle form an indefinite length of paper, and the development of the larger and more powerful Foudrinier machine in England a few years later, the ancient traditions of a hand craft were transformed into a modern industry.

In the Foudrinier method, paper is manufactured by pouring pulp onto a moving, continuous wire screen. Most of the water immediately drains away. The newly formed web of fibers is transferred onto a felt which supports the paper as it passes under a series of rollers which press out more water. The paper then passes over heated drying cylinders. Calender rolls give the paper a smooth surface finish before it is finally wound up in a roll.

When the first factory for machine-made paper was established in Shanghai, China, in 1892, the story of paper returned to the place where it had begun almost 2,000 years earlier.

A 19th-century papermaking machine with three drying cylinders, capable of forming an indefinite length of paper and imitating mechanically all the operations that are performed in the hand process.

A CONTEMPORARY ART MEDIUM

Artists have always concerned themselves with the characteristics of paper – its weight, texture, and tone – as an integral part of their work. Today, however, artists have recognized the potential of the material as a medium in itself. The tremendous versatility of paper pulp accounts for much of its current popularity. It can be cast for sculpture; manipulated by hand like clay; burnished like metal or embellished with gold and silver leaf; vacuum-formed; sprayed, poured, daubed or spattered like paint; spun into thread and woven as textile; sawn like wood or carved like stone.

Others are interested in paper as a medium for opposite reasons. Paper can quite simply be itself, and the process of making it need not imitate other techniques. The naturally structured procedures of making paper by hand have engaged and inspired many artists, regardless of their background discipline. The gathering of the raw materials, the preparation and beating of the fibers, the decisions that determine the final appearance of the sheet – its color, surface texture, translucency, softness or strength; the almost instantaneous moment when the pulp settles on the surface of the mold; even the drying of paper offers a further possibility of allowing the paper to assert its own original expression.

Alan Shields: Rain Dance Route. Relief screenprint, woodcut (left). As an innovative printmaker, Shields came to regard paper as a medium as well as a support. He has constructed brilliantly colored lattices with his prints.

David Hockney: Paper Pool (No. 31) (right). David Hockney made a group of paper works of swimming pools in collaboration with Ken Tyler at Tyler Graphics, New York, in 1978.

Julia Manheim: Fallen Idol (left). For this sculpture, Manheim used a combination of Soviet Crisis newspapers and steel wire.

CHAPTER

Getting Started

Japanese "rakusui" paper made with leaves embedded between the layers (right).

During nagashi-zuki sheetforming (right), the papermaker gradually laminates a thin sheet of paper by repeatedly sloshing the vat mixture across the mold's surface.

Papermaking – East and West

A European vatman (above) dips the mold into the vat once and pulls it out, laden with pulp, before quickly shaking it from side to side and back to front to even the sheet.

Paper is made from the cellulose found in plant fibers, which are literally beaten to a pulp and dispersed in water. Forming the pulp into a sheet of paper is accomplished with a "mold" and "deckle." The mold is a rectangular wooden frame over which a porous screen or cloth (the mold cover) is stretched. The deckle is an uncovered frame which fits onto the mold. It forms a raised edge around the screen that prevents the pulp from running off the mold during sheetforming. A sheet of paper is formed by dipping the mold and deckle into a vat containing the pulp mixture and picking up a small quantity of pulp on the surface of the mold. As the mold and deckle are lifted out of the vat, the water drains through the screen, and a layer of interlaced fibers settles on the mold – a wet sheet of paper is formed.

There are two distinct methods of hand papermaking, one Western and one Eastern. Both are centuries old. Both have remained essentially unchanged. The Japanese defined the difference between the two methods, calling the Western technique *tame-zuki.* This translates roughly as "the fill and hold way to make paper." Traditional Japanese practice, which characterizes the Eastern method of making paper, is called *nagashi-zuki.* This translates roughly as "the flowing or sloshing way to make paper." Each term describes the method used to form a sheet of paper. A third traditional approach, based on ancient Chinese practice, is called the "floating mold" method. It is still used by papermakers in Nepal, Bhutan, and other Himalayan countries.

Both Eastern and Western techniques can appear complicated, but the basic workings are not difficult to grasp. The process of hand papermaking has been adapted throughout its history to suit local conditions of manufacture. With a little practice, and perhaps by modifying a particular technique to suit your own skills and requirements, you can start to explore the dramatic range of visual and tactile qualities of handmade paper.

BASIC EQUIPMENT

Never let the lack of conventional papermaking equipment prevent you from getting started. With a little improvisation, readily obtainable tools and materials can be adapted to suit your purpose. Some of the earliest methods of making paper, which might seem primitive when compared to contemporary processes, resulted in the most beautiful and enduring examples of handmade paper. As you progress from the basic techniques to more advanced procedures, you will be able to determine the future scale and focus of your papermaking requirements and invest in equipment and supplies accordingly.

To make your own paper you will need the following:

- a mold and deckle
- a source of fiber
- a device for beating your fiber into a pulp
- a clean water supply
- a vat or container for holding the diluted pulp.

If you already have a blender and plastic sink-liner bowl, you need only construct a simple rectangular wooden frame and attach a screen-wire surface to it to begin a simple papermaking operation.

WATER SUPPLY

A supply of fresh water is essential for any papermaking venture. Each step of the process — soaking the fiber, rinsing, cooking and beating it, through to sheet-forming — involves the use of water. You should, therefore, equip your work space to deal with a certain amount of spillage. Tap water, whether hard or soft, should be clean and, wherever possible, free of contaminants. If you want your paper to last a long time, it is best to determine the composition of your water supply (your local municipal water authority will be able to tell you what minerals are present), and to purify it if necessary. You can buy an inexpensive test kit (used for aquariums) to check the acidity and alkalinity of the water. Mineral and organic residues tend to migrate to the surface of the paper as it dries and will precipitate quite visible staining. If you are making paper with an anticipated short life, you may not need to worry about purifying your water system.

THE VAT

The container which holds the prepared pulp is known as a vat. A serviceable vat can be adapted from virtually any container — for example, a plastic sink bowl, cat litter box, or food storage tray. Even a kitchen sink with a suitable drainhole cover will do. Sheet plastic can be used to line a temporary unprimed vat. The vat should be a few inches deeper than the shorter side of your mold and deckle. Its inside dimensions should allow room to maneuver the mold and deckle, with enough space for your hands on each side.

Basic equipment for making your own paper includes a mold and deckle, a vat, a liquidizer, a strainer, couching felts, and pressing boards.

The frame for nagashi-zuki sheet-forming consists of a ribbed mold; a flexible screen surface, or su; *and the deckle which is hinged to the mold.*

Mold and deckle

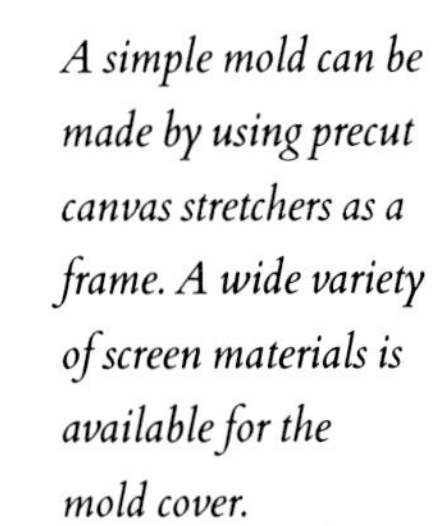

A simple mold can be made by using precut canvas stretchers as a frame. A wide variety of screen materials is available for the mold cover.

The mold and deckle is the papermaker's chief tool. Although its construction differs according to each particular sheet-forming method, its function remains the same: to allow the water to drain away from a thin deposit of pulp that then becomes a piece of paper.

Japanese papermaking molds are called *sugeta.* The *su* is a removable, flexible screen, usually made of thin, closely spaced, parallel strips of bamboo held together with fine silk threads. The *keta* is a lightweight, hinged wooden frame that holds the su in place. Ribs on the lower frame support the su and, on larger keta, two handles are fitted across the upper frame.

A traditional Western mold is made to withstand the weight of a larger amount of pulp and water than its Oriental counterpart, and is therefore a sturdier construction. There are two types of Western molds: "laid" and "wove." A Western laid mold resembles the Japanese sugeta, but the screen cover is permanently attached to the mold frame and is made out of thin wire rods. The closely spaced, horizontal (laid) wires are similarly held together with more widely spaced perpendicular chain wires. A system of wooden ribs also supports the mold cover, but the deckle is not hinged to the mold. Paper made on a laid mold is called laid paper. It can be distinguished by the ribbed pattern of horizontal and vertical lines caused by the laid surface of the mold.

A wove mold refers to any mold with a woven mesh surface. The earliest mold was probably a simple frame with a coarsely woven cloth stretched across it. A Western wove mold covering now consists of a finely woven bronze wire cloth. The wove paper it produces usually has a smoother surface appearance than laid paper.

MAKING A MOLD AND DECKLE

If you have some basic carpentry skills, you can make a mold and deckle using the method illustrated on p. 22. However, a simple alternative is to use cut-to-size canvas stretchers. They are usually sold in pairs of equal length. You will need two identical sets of two matching pairs for a complete mold and deckle. To construct the lower mold, fit the pieces together and check to see that the corners are square. Fasten the corners with brass brads or small screws. Rub down any rough or uneven surfaces with fine sandpaper. Assemble the pieces for the deckle in the same way as the mold. Waterproof with sealer.

The deckle determines the shape and thickness of the sheet and can be made accordingly. You can make a sheet of paper without a deckle, but the pulp will run off the edges of the mold and result in a less well-formed sheet. Pulp which slips under the deckle during sheetforming is responsible for the characteristic "deckle" edge of handmade paper. An adhesive closed cell foam strip, such as draft excluder or weather stripping, can be taped around the edges of the underside of the deckle to give a tighter fit on the mold.

The mold must be covered with a screen surface. Suitable materials include glass curtain fabric, silk-screen mesh, or window screen. Hardware cloth can be used to support a finer mesh fabric such as nylon window screen, which otherwise tends to sag. Heat-shrinking polyester mesh and brass screening can be ordered from papermaking suppliers. The screen mesh should be fine enough to allow the fibers to remain on the surface and the water to drain through. If you choose a nylon screen mesh, such as curtain fabric, which tends to stretch when it is wet, you will need to wet it before stretching so that it dries taut.

A European wove mold made from mahogany with a fine yet strong phosphor-bronze screen surface. A division stick for the deckle allows two sheets to be made at a time.

PROJECT

Making a mold and deckle

1 To construct the lower mold, fit the lengths of wood together. Use simple abutment, lap or dovetail joints. The correct inside measurement of the frame should be $8\frac{1}{4} \times 11\frac{3}{4}$in. ($21 \times 30$cm).

2 Glue the corner joints with a waterproof adhesive and nail with brass brad nails. Brass L-shaped corner plates with flathead brass screws will strengthen the corners.

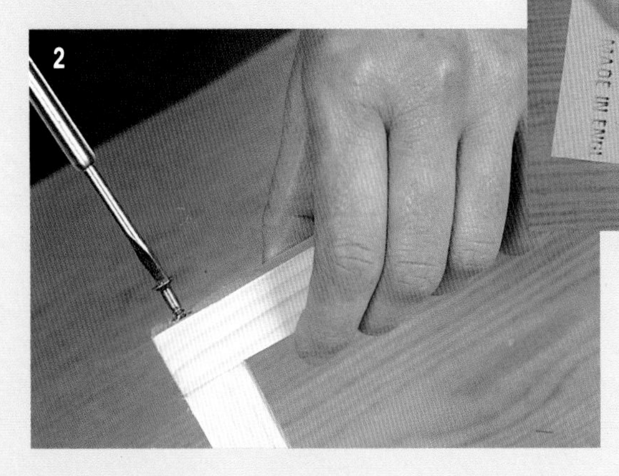

3 Rub down any rough or uneven surfaces with fine sandpaper.

YOU WILL NEED

- *8 lengths of wooden slats*
- *waterproof adhesive*
- *brass brad nails or small screws*
- *brass L-shaped corner plates (optional)*
- *waterproof adhesive*
- *waterproof sealer or linseed oil*
- *screen wire*
- *stainless steel staples or brass tacks*
- *fine sandpaper*

4 The mold must be covered with a screen surface. It must be stretched taut across the mold. Cut screen wire the same size as the mold or just within the outside edges. Stretch glass curtain fabric over and around the outside edges before it is stapled. Using staples or tacks, work from the middle of each opposite side toward the corner edges.

5 Apply waterproof tape around the edges of the screen over the staples to prevent pulp from being caught under the screen wire.

6 Assemble the deckle pieces in the same way as the mold. Use a penetrating waterproofing sealer or a light coat of linseed oil on both the mold and the deckle.

7 The deckle fits neatly over the finished mold, ready for use.

The long staple fibers of the cotton plant are removed by a process called "ginning" and used to make cloth. Cotton rags produce a very strong, durable paper.

Sources of Fiber

Leaf fibers with a strong outer skin must be scraped, or decorticated, to remove the skin and release the inner fibers.

Cellulose fiber, the main ingredient of paper, is available in a number of different forms. All living plants contain cellulose, but some yield a higher percentage of usable fiber than others. Some fibers are long and slender, while others are shorter and thicker, and there may be different length fibers within the same plant.

Paper can be made from almost any plant, but those with a high content of long cellulose fibers generally provide the best source. It is also possible to use fiber that has previously been processed and used for another purpose. Your choice of fiber will be determined by the kind of paper you want to make and the equipment you have for processing it.

Raw fibers for papermaking are classified according to their location in the plant. Bast or inner bark fibers, such as hemp, flax and kozo, contain some of the longest fibers for papermaking. Leaf-stem fibers, such as abaca, and leaf fibers from plants such as sisal and yucca offer a comparatively shorter range than bast fibers. Grass fibers include such plants as bamboo, bagasse (sugarcane stalk), and rice straw. Grass fibers produce the shortest and most brittle fiber for papermaking, and the amount of usable fiber is relatively low.

A further category of fibers are attached to the covering around the seeds of certain plants. Cotton and kapok are familiar examples of "seed-hair" fibers. Finally, there are fibers from both softwood and hardwood trees, which are used for industrial papermaking.

PREPARED FIBERS

Fibres used in textile- and basketry-making are often suitable for making paper. Cloth made from flax and cotton

Dry bundles of Japanese bast fibers (below left): kozo, gampi, and mitsumata. The bark needs to be soaked overnight to soften it prior to cooking. Harvesting flax fibers (left) from which linen cloth is produced. Flax can be used to make exceptionally strong, translucent paper. Cotton (below right) is a strong, versatile fiber which provides an exceptionally pure form of cellulose.

was the traditional source of fiber for Western papermaking. "Half-stuff" (partially prepared cloth scraps from the textile industry) is perhaps a more common source today. Processed fibers, such as raffia, seagrass, jute and sisal, are often sold by craft and weaving suppliers for textile purposes. Other partially processed fibers include cotton linter, which comes in several grades, and abaca. These are available in the form of dry, compressed sheets.

An alternative source worth considering is recycled paper pulp. Paper has a long history as a recycled material, and today many artists believe in using waste papers as part of a wider environmental policy. Industrial and commercial papers provide a readily available source, but they should not be used indiscriminately. Papers with either a heavily coated or glossy surface, or those treated with a hard surface size (which renders them less absorbent) are difficult to break down without prolonged soaking or cooking. Poor quality, short-life papers, which contain a large percentage of wood pulp and are highly acidic, will produce an inferior recycled paper. If high standards are required, the best papers for recycling are quality handmade or moldmade rag papers.

Rice straw has long been used as a material for papermaking in China. The piled straw has been treated with lime to release the fiber.

Preparing the fiber

Japanese bast fibers are steamed in wooden containers. Once the stems are cool enough to handle, the loosened bark is peeled off the woody core.

Most fibers need some kind of treatment before they can be used for making paper. This usually involves cooking the fiber in an alkaline solution and, sometimes, a pre-cooking stage, depending on the nature and form of the fiber.

The methods used to prepare linen and cotton rags for early European papermaking were quite extensive. Before the cloth could be used, the effects of spinning the cellulose fibers into threads and weaving the threads into cloth had to be undone. This was a time-consuming and labor-intensive task. The rags were first sorted and graded according to fiber, color, and quality, and any seams or buttons removed. After sorting, any loose dirt was removed, and the rags were cut into smaller pieces. The rags were then put through a process of fermentation, which began to loosen and soften the fibers for beating. This was followed by careful and thorough washing. Faster cooking and bleaching methods began to replace these traditional techniques during the early 19th century.

RAW FIBER FERMENTATION

Raw fibers often require fermenting (also called retting) before, or in place of, cooking to render them suitable for papermaking. Fermentation can be achieved by soaking the fiber in an alkaline solution or just by leaving the fiber to stand in water in a warm place until bacterial action begins to loosen the fiber. This traditional technique is still used by the Chinese to prepare bamboo for papermaking.

The preparation of kozo, which represents 90 percent of the bast fiber used in traditional Japanese papermaking, also involves several preliminary stages.

Kozo trees, a variety of mulberry, are usually harvested after two years of growth. The stems, which are about ½in. (15mm) in diameter, are cut into even lengths, bundled, and steamed in a closed container for two hours. The steam causes the bark to shrink back from the wood core and can then be easily stripped away. The stripped fiber is hung to dry and stored for future use.

Before it can be used for papermaking, the dried fiber has to be softened, and it is left to soak in cold running water, usually a nearby stream, for several hours. Quite often, the bundles of soaking fiber are trampled under foot, which loosens the black outer layer. The fiber is then taken to a dry work area, and the black outer bark is scraped off the damp strands with a knife. If a fine white paper is desired, the secondary green bark is also removed, leaving only the white inner bark for papermaking. The cleaned fiber is then washed to remove any specks of bark. It is either dried and stored again or taken directly to be cooked.

COOKING

After preparation, raw fibers usually need to be cooked before they can be properly beaten into a pulp. The aim of cooking is to liberate and purify the fibers and dissolve unwanted parts of the plant. Lignin (the intercellular "glue" which binds the fibers together as the plant grows), for example, is undesirable because it rejects water and impedes the bonding of the cellulose fibers, which are attracted to each other in water. Other extraneous material, such as sugars, starches, waxes and gums (called "extractives"), must also be removed. Their presence will contaminate the pulp and can cause the sheet to discolor and deteriorate with age.

The objectives of cooking are best achieved using a mild alkaline solution. In the past, Japanese papermakers used a potash made by leaching water through wood ash (potassium carbonate), and sun-bleaching was often used to lighten the color of the fiber before beating. The bark is cooked for several hours, depending on the age and quality of the fiber and the kind of paper being made. After cooking, it is washed thoroughly to rinse the alkali from the fiber. Each strand is then examined for any remaining impurities and carefully picked clean.

Sorting, grading, and cutting rags, which arrived at the mill jumbled together in all sorts of conditions, was an unenviable task usually carried out by women.

COLLECTING AND COOKING

Finding local plant sources is a rewarding aspect of making your own paper. You might well discover in your area relatives of plants that have already been identified as a proven source of papermaking fiber.

Most plant fibers are harvested in the late summer or early fall, or at the end of the growth cycle, when the cellulose content has reached its peak and before the plant has begun to dry out. Just as the fiber from each plant species differs, fiber from a single plant can also vary according to seasonal climate, conditions of growth, age, the time when it is harvested, and the method used to process it.

If you decide on a particular plant, try picking it at different times of the year. Use it green or allow it to dry before cooking it to produce a subtle range of texture and color variations in your finished paper.

HOW MUCH TO COLLECT

The amount you collect will depend on the availability of the fiber and how

Weighted bundles of bamboo are left to soak as part of the lengthy fermentation process required to prepare the fibers for papermaking.

much paper you want to make. Some fibers contain a higher volume of fiber than others. Yucca, for example, is easy to harvest and the leaves of the plant yield a high proportion of fiber, but some reeds which would seem to provide a strong fiber yield only small quantities of short fibers after cooking. Surprisingly large quantities of raw material are needed to produce a useful amount of pulp from plants with a lower yield. It is better to start with a small amount of fiber, and find out how successful it is, before you collect in bulk.

COOKING TIMES

Cooking times will vary according to your choice of chemical. The chemicals used in the cooking process affect the characteristics of the finished paper. A strong caustic solution, or prolonged cooking, can result in damaged fibers and a weaker paper. Some fibers take less time to cook than others, and tougher fibers may require longer cooking.

If a fiber is not sufficiently cooked after two hours, it may be best to cook it a second time. Strain and rinse the fibers, and make a fresh alkaline solution for a further two-hour cook. If a fiber fails to cook enough after four or five hours, it is probably advisable to try fermenting it first.

It is important to note that herbaceous bast fibers, and some of the tougher leaf and grass fibers, usually need a fair amount of preparation. Some are resistant to cooking and, in addition, need prolonged beating, either by hand or machine, to produce a suitable material for papermaking. Such fibers are

METHYLCELLULOSE

Methylcellulose comes in powder form and is water soluble. It can be used as an archival, mild glue and is often added to short fiber pulps to increase bonding for casting. Make a stock solution of methylcellulose with 16oz.(454ml) of water and 8oz.(227g) of powder. Add the powder to the water gradually, stirring the solution as you do so. Leave until the powder has dissolved completely. Dilute the stock solution as required.

FORMATION AID

Formation aid, also called *neri* or synthetic *tororo-aoi*, is available from papermaking suppliers and comes in a powder form. The powder should be mixed with cold water using 2tsp.(4g) of powder to 1 gallon (4 liters) of water. Fill your mixing container with water and add the powder slowly, stirring it as you go. Leave the solution for several hours, stirring occasionally, to make sure that the powder dissolves completely.

Dry, field-retted leaf fibers can be harvested late in the fall and will not need to be scraped before cooking. Cut the leaves into small pieces and soak them in clear water for 24 hours.

better prepared by using fermentation techniques (see p. 25).

THE CHEMICALS TO USE

Soda ash (sodium carbonate) is considered a standard cooking chemical. It is a safer and more convenient choice than caustic soda (also known as lye). Soda ash, or washing soda, can be ordered from papermaking suppliers and most retail chemical suppliers. The following directions are for cooking plant fibers with pure soda ash.

COOKING INSTRUCTIONS

Having identified and collected your plant material, cut the stem or leaves into ½in. (12mm) pieces. Remove any nodes, or joints, and discolored parts. Pre-soaking the fiber can reduce cooking time. Dried plants should be weighed and soaked for 24 hours prior to cooking. Some plants (those with hollow stems, for example) may need to be crushed with a wooden mallet or rolling pin before soaking. Fresh plant material should be weighed and soaked overnight.

The alkaline (soda ash) ratio can be determined according to either the dry weight of the fiber concentration, or the volume of water used to cook the fibers. For the dry weight method, an average of 20 percent of the dry fiber weight is recommended — 20 percent of 16oz. (450g) of fiber, for example, is 3¼oz. (90g). However, since the volume of water determines the strength of the solution, this is a more accurate method of formulating the alkaline ratio. If you use this approach, begin with ½oz. (15g) of soda ash to every quart (liter) of water. You can raise the concentration to 1oz. (30g) per quart (liter), depending on how difficult the fiber is to cook. The pH value of the cooking solution should be between 10 and 11 to eliminate undesirable noncellulose constituents, yet leave the cellulose unharmed. Use a pH test strip to check the balance.

Fill a stainless steel or enamel cooking pan with enough cold water to cover the fibers. Measure the required amount of soda ash and dissolve it in water. Add this solution to the water in the pan. Stir the solution well and bring it to a boil. Add the fiber when the solution begins to boil.

Bring the fiber and cooking solution back to a boil and then reduce the heat to simmer. Time the cooking from this moment. Cover the pan and cook for 2-3 hours in a well-ventilated area, stirring the fiber every 30 minutes. Check the fiber at the same time by removing a small amount and rinsing it. Then try to pull it apart along the grain. The fibers are cooked when they separate easily with a gentle tug.

Let the cooking solution cool and then strain off all the used liquid,

Once the fibers are cooked and have been drained, rinse them thoroughly under clear running water making sure that all the fibers are cleaned of contaminants.

preferably using an outside drain. Rinse the fiber thoroughly until the water runs clear. Use a deep strainer with a fine nylon or non-corrosive mesh, or a colander lined with a fine mesh fabric. Always wear rubber gloves when draining and rinsing the fibers. It is essential to remove all traces of the exhausted chemicals and any vestiges of lignins and other nonfibrous constituents. The simplest method is by repeatedly washing the drained fiber in a bucket, using several changes of fresh water. This should reduce the pH of the fiber to almost neutral (pH 7). You can check this balance by dispersing a small amount of fiber in a container of distilled water and taking a reading with a pH test strip. (Available from library or aquarium suppliers.)

A 3 percent solution of hydrogen peroxide can be used to lighten fibers with less risk of degradation than from other chemical bleaching agents. Before bleaching (right) and after (below).

SAFETY CONCERNS

- **Wear gloves and eye goggles when mixing chemical solutions.**
- **Never use aluminum containers or stirring utensils with alkalis.**
- **Always add the alkali to the water and never the other way around.**
- **Avoid skin contact with solutions and work in a well-ventilated area.**
- **Always read safety precautions and follow manufacturer's instructions.**

BLEACHING

Plant fibers have their own characteristic colors and rarely need bleaching. Several traditional methods of fiber preparation, however, have the effect of lightening the fibers, which can enhance the qualities of the finished paper. In Japan, for example, hanging stripped fibers outside to dry exposes them to the natural bleaching effect of sunlight; and in Europe, the routine procedure of washing during beating was once a common method of lightening rag fibers.

The introduction of chlorine and its use as a bleaching agent at the end of the 18th century quickly replaced older, more time-consuming methods of whitening rags without chemicals. Chlorine can, however, seriously degrade cellulose fibers and alter the pH balance of paper, causing it to become yellow and brittle with age. If you do wish to lighten the natural color of your fiber, hydrogen peroxide is the safest chemical bleaching agent to use. Steeping the fibers in a dilute solution for several days after cooking will cause them to brighten gradually. Optimum bleaching takes place in a solution with a pH balance of between 9.0 and 9.5. Hydrogen peroxide will not leave a harmful residue on the fiber, but you must wash the fibers thoroughly after bleaching. The pH should be close to neutral.

STORING FIBER

Dried plant fibers can be stored indefinitely, but make sure that they are kept dry and do not mildew. Cooked and washed fibers may be covered with water and kept in a refrigerator for a week, and sometimes longer. Check the fibers at regular intervals to see that they have not spoiled. Cooked and washed fibers should not be dried for future use, but they can be frozen.

A set of water-powered stampers at the Richard de Bas paper mill in France. These were used for pounding and refining rags in water-filled troughs.

Beating and pulp preparation

A Hollander beater filled with pulp. Hollander beaters cut and lacerate fibers, producing a different quality pulp from that produced by the rubbing and fraying of stampers.

How you beat your fiber will have a significant effect on the quality and characteristics of your finished sheet of paper. The aim of beating is twofold: to hydrate and fibrillate the fiber. Hydration refers to the alteration of the cellulose fibers as they absorb water during the beating process. Fibrillation refers to the bruising of the fiber walls and unraveling of the fibrils – the smaller fiber components of the fiber. Both effects promote the chemical bonding (known as hydrogen bonding) between the cellulose fibers that is necessary to make a sheet of paper.

Shortening the length of the fibers through cutting is an optional effect of beating. A certain amount of cutting can be useful, particularly with longer fibers, which may clump together to create a wild or unevenly formed sheet of paper. Too much cutting, however, will produce a paper without suppleness or strength.

The early European method of beating prepared rags into pulp was by using large, waterpowered trip hammers, or stampers. These were replaced during the late 17th century by the Hollander beater, which was not only faster, but could produce pulp from unprepared cloth.

Today there are several types of Hollander beater, but all of them work on the same principle: the pulp circulates around an oval-shaped tub, which is divided in the middle. A powered roller with fixed metal bars across it (rather like the paddles on a paddle steamer) is situated midway down one side. As the roller rotates, it pushes the pulp down one side of the tub, where it passes between the roller and a bedplate, which is anchored on the tub floor beneath the roller. The pulp is thrown up and over the backfall, around the back turn, up the opposite side, around the next turn, and so on. The fibers are continually beaten against the bedplate by the roller bars until they are turned into a pulp.

Beating methods and beating times are determined by the nature of the fiber and the type of paper required. In Japanese papermaking, relatively little beating is needed, and beating continues only until the fibers are loosened. This is because the bast fiber contains a high percentage of hemicellulose, which is similar to cellulose in composition but more easily plasticized.

Plant fibers can be beaten by hand using a wooden beating stick (left), as in this illustration from Kamisuki Chohoki *(above).*

BEATING BY HAND

Hand beating is the simplest and most direct method of turning plant fibers into a suitable pulp for making paper. The original Chinese method of beating fragments of cloth or bark into a pulp was by pounding with a pestle in a mortar. You can adapt this method for most plant fibers, using an implement such as a wooden mallet, a baseball bat, or the flat edge of a piece of hardwood, and beating the fiber on a solid piece of hardwood or granite block. Even the flat end of an ax-handle in a strong bucket will work. Soak any wooden implement and wet the board before you start beating to help prevent the wood from splitting (for this reason, do not use an ordinary chopping board).

If you adapt a pestle or mortar from a smooth stone or river rock, make sure that pieces do not chip or flake off into your fibers while you are beating them. Some fibers have a tendency to splatter as they are beaten. If you are beating on a flat surface, you may wish to make a splatter guard around your beating area. It is advisable to wear goggles.

Take a small quantity of cooked fibers, squeeze out the excess water, and place on a sturdy beating surface. Beat the fiber by moving systematically across the pile. The fiber will begin to expand under the force of each blow. As the pile spreads out across the beating surface, stop and fold the edges in toward the center. Then turn the mass of fiber upside down and repeat the procedure. If the fibers appear to be drying out while you are beating, sprinkle a small amount of water on them so that they can properly hydrate and separate.

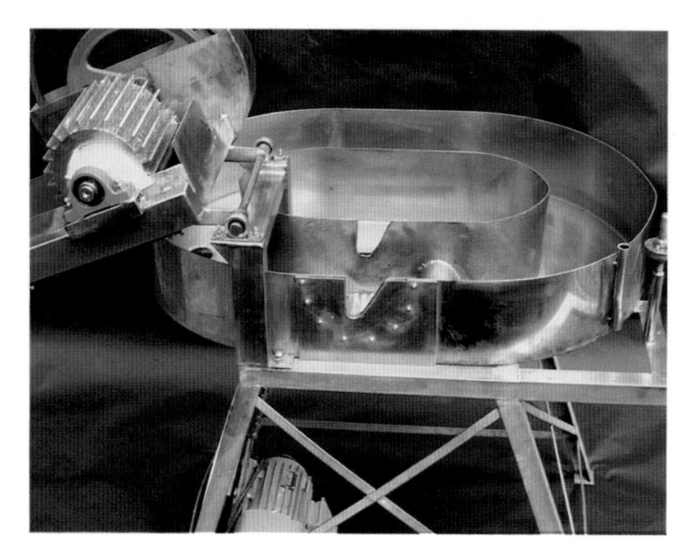

A Hollander beater (designed by Peter Gentenaar) folded open to reveal the roll, and bedplate. This model is made for processing long fibers such as flax.

TEST FOR READINESS

To determine whether the fiber is sufficiently well beaten, add a small amount to a jar of fresh water and shake vigorously. If there are any large strands or clumps of fiber, continue beating.

USING A BLENDER

A kitchen blender offers a quick and easy alternative way of preparing some plant fibers, although you may wish to continue the process of preparation by hand methods and reserve the blender for recycling paper. Place a small handful of prepared fiber in the blender, and fill it with water to a level approximately 2in. (5cm) from the top. Run the blender for an initial 10-15 seconds. Watch carefully to make sure that the fibers do not become entangled in the blades, and listen for any sound of strain on the motor. Make sure that water does not enter the motor. If your blender has variable speeds, start with a lower one and increase the level according to the stage of fiber separation. You will put less strain on the motor by running it in a series of short bursts than a single longer one. Remove the lid and check the fiber between each burst. Never leave the blender running unattended or allow it to overheat.

Remember that a blender, because of its purpose, will separate and shorten the fibers by cutting them. A blender cannot completely hydrate or fibrillate the fibers. Each time you "recycle," or liquidize, a fiber, its length is shortened and the finished sheet will be consequently weaker. Hand beating rarely results in over-shortened fibers.

RECYCLING PAPER IN A BLENDER

Whatever type of recycled paper you choose (see below), make sure that any pieces of tape, gummed edges, staples, or paper clips have been removed. Tear or cut the paper into 1in. (2.5cm) squares or smaller pieces. Soak them in water overnight. If the paper is not absorbent, you may need to soak it for longer or use boiling water. Rinse the pieces after soaking.

Waste papers – especially rag paper which has been heavily sized, or paper which has already been used for writing, drawing, or printing – can be cut into pieces and then cooked in a mild alkaline solution to clean and soften the

Plant fibers can be beaten by hand using a small, heavy pestle and mortar, or a wooden mallet on a broad solid surface.

Assorted axial and radial flow propellors and mixers can be used to rehydrate partially prepared pulps, as well as to blend colors and other additives into your pulp.

fibers. Rinse the pieces again before processing in the blender.

If you are using an acidic source material, such as coated magazine paper or newsprint, and are concerned about the permanence of your paper, you will need to use calcium carbonate (a pure form of limestone) as an alkaline buffer. Soak the pieces overnight in a solution of 1¾oz. (50g) of calcium carbonate per quart (liter) of water.

To prepare your pulp, take a small handful of paper pieces and drop them, one at a time, into the blender filled two-thirds full with water. Blending times will vary according to how smooth or textured you wish the paper to be. Liquidize printed paper for a relatively shorter time if you want to preserve parts of the image or text.

Other mixing devices, such as a Whiz-mixer (a propellor-like blade on the end of a long motor-driven shaft), can be used to make larger quantities of pulp from partially-prepared fibers in sheet form.

STORING BEATEN PULP

Leftover beaten pulp can be drained, stored in a covered container, and refrigerated. If the weather is cool, you may be able to leave the pulp in a bucket for several days. Cover the bucket with fine screen or a porous cloth to stop insects or

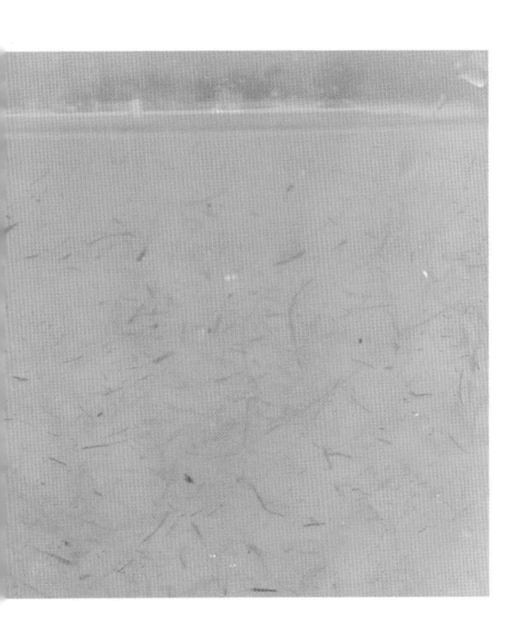

Small batches of plant fibers can be processed using a blender (right and below). Do not liquify completely, or you will have weak, brittle paper.

Beat the fibers (above) until they are reduced to an equal fineness when shaken in a jar of clear water.

dust from settling on the pulp. Like any other vegetable material, it will spoil if left for too long.

You can leave leftover beaten pulp to air dry, once it has been drained. Use a fine strainer or net bag, and squeeze as much water as you can out of the pulp. When you need to use the pulp again, soak it in water until it softens. You may then need to remix it to eliminate any undissolved lumps. However, once the pulp has dried, it will never reabsorb the same amount of water, and may produce a weaker paper.

COLORING AND OTHER ADDITIVES

A wide range of natural substances – including berries, barks, plant juices, soot, saffron, mollusks, madder, woad, and indigo – were used to color early papers. Various methods were used to apply the color, such as brushing it onto the surface of the paper or dipping the sheet into a dye bath.

Early European handmade papers reflected the color and quality of the rags from which they were made. The principal colors were shades of white, brown, and blue. The best rags were selected for white paper. This was available in several grades and was used for writing and printing. Blue and brown rags, for the most part, produced a range of coarser, but strong, serviceable papers, which were used for wrapping and for decorative papers and wallpapers.

By the 18th century, the addition of colorants to the pulp – following the principles of textile dyeing – had

A subtle range of papers can be produced using the natural colors of different plant fibers.

Water-dispersed pigments are made specifically for paper pulp to give rich and vibrant colors.

extended the variety of colored paper. Dyes are water-soluble coloring agents. They usually penetrate the structure of the fibers in order to become attached to them. Dyes have a natural affinity for cellulose. The most frequently used include direct dyes (organic dyes usually derived from coal tars); fiber-reactive or procion dyes (which form a chemical bond with the fibers in an alkaline solution and are often used in the textile industry); and natural dyes derived from plant sources.

Few dyes are entirely colorfast, lightfast, or runproof when used alone, and you need to be careful over selecting the most suitable dyestuff. Dyes which need an acid mordant (color-bonding substance) to fix them onto cellulose fiber should be avoided. A book dealing specifically with the use of textile and natural dyes will advise you on preparing the correct dyestuff for your choice of papermaking fiber.

Pigments are insoluble, finely ground particles which either coat the fibers or fill the spaces between them. The only truly lightfast colorants, they produce clear, bright shades of color and are chemically inactive. However, pigments have little affinity for cellulose and must be held onto the fiber by means of a binder. Pigment and binder are usually added to the pulp during the final stage of the beating process. A choice of aqueous-dispersed pigments (dispersed in a concentrated form in water), specifically formulated for coloring paper pulp, are available together with the necessary retention agent, from papermaking suppliers, and come with full instructions for use. The retention agent is designed to help bond the dye to the fibers in place of an acid mordant.

BRIGHTENERS AND FILLERS

Certain additives, such as calcium carbonate, kaolin (China clay), and titanium dioxide, can be used to improve the opacity and brightness of paper. They usually come in fine powder form and are added to the pulp at the end of the beating cycle. Like pigments, they become entrapped between the fibers and can be used as a "filler" to improve the surface smoothness of a sheet for printing, or to reduce shrinkage in three-dimensional paper techniques.

A selection of Japanese colored papers.

A shallow pool of water is used as a vat in Nepal. A sheet of paper is formed by pouring the pulp into a mold floating on the water.

Vats

Early Western vats resemble large wine casks with metal hoops around the wooden staves.

The vat or tub which contains the prepared pulp for papermaking has changed over time, and according to the method of sheetforming and the shape of mold being used.

One of the oldest vats in Eastern papermaking is a shallow rectangular trench or well dug in the ground and filled with water. The mold (a wooden frame with a woven material stretched across the bottom) is floated on the water, which fills the bottom of the frame. The papermaker squats beside the vat and forms a sheet by pouring a measured amount of pulp into the mold, leveling the pulp by deft movements of the hand before carefully lifting the mold from the vat.

Most Eastern methods now employ a rectangular wooden vat raised to a convenient standing height. This is filled with water to which the beaten fibers are added. The vat is large enough to accommodate the mold and the action needed to scoop up the pulp and form a sheet of paper. Vat attachments for traditional Japanese vats include a *maze* (mixer): a removable rake-like device suspended above the vat for distributing the fibers evenly throughout the vat, mold support sticks, and a mixing stick.

In the earliest depictions of European paper mills, the vat is often a large, round wooden tub (although the molds were rectangular), and was probably made from a huge wine cask. Later illustrations show the "pistolet," or heating device, which warmed the water in the vat, and a bridge or platform which extended across the vat and supported the "horn" against which the mold was placed to drain prior to the couching of the freshly formed sheet. Production vats were later made of stone or iron (lined with lead to prevent rust). Their design included a "knotter," which filtered the knots or other impurities from the pulp, or stock, before it entered the vat; and a "hog," a rotating paddle device at the bottom of the vat which kept the stock in suspension.

Wood (pine and cedar) lined with copper or stainless steel sheeting, concrete (sometimes lined with ceramic tiles), and marine fiberglass are used to build contemporary production vats for Japanese and Western papermaking. They usually follow a rectangular design, with the front wall slanting toward the bottom, and include many of the traditional vat accessories.

FILLING THE VAT

To make a sheet of paper, you must first dilute your prepared pulp in a vat of fresh water. The thickness of a sheet is determined by the proportion of pulp to water. A good ratio to begin with is one blender jar of pulp to four of water. You can either half-fill the vat with water and add a blender jar full of pulp, stirring in extra pulp until you reach the desired consistency, or you can pour the pulp from the blender into the vat first, adding water to dilute it. When you are filling the vat, leave a distance of about 3in. (8cm) from the top; otherwise, the pulp and water are liable to splash over the sides as you make a sheet.

Each time you make a sheet, you remove a certain amount of fiber from the vat solution. You will need to replenish the vat with more pulp as the mixture becomes thinner. Keep a bucket of prepared pulp near your vat and, draining a measure at a time through a plastic strainer, add extra pulp to the vat as required.

Most pulp has a tendency to settle at the bottom of the vat. Each time you make a sheet, stir the vat mixture gently to distribute the pulp evenly throughout the vat. Wet the surface of your mold before you make your first sheet, and then position the deckle on top of it.

You will have to experiment with the ratio of fiber to water, but you will soon be able to judge the consistency, and vary it according to how thick or thin you want the paper to be.

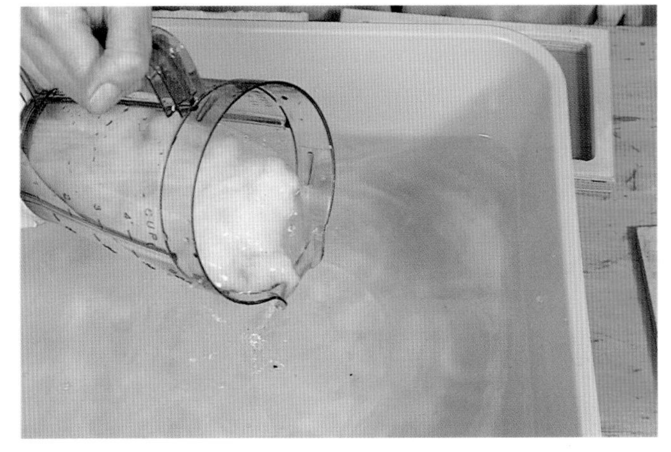

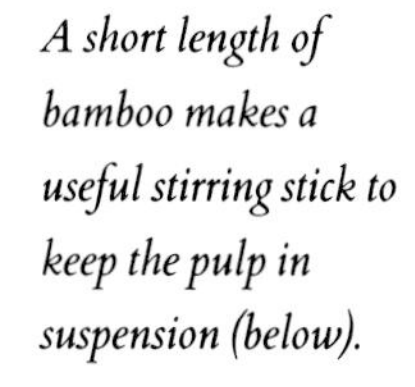

A short length of bamboo makes a useful stirring stick to keep the pulp in suspension (below).

The papermaker in Japan works alone, compared to the operation of a vatman and coucher in the West.

Sheetforming and couching

After the paper has been formed and the deckle removed, the coucher transfers the sheet onto a felt. A slightly curved couching surface helps to facilitate the transfer.

The standard Western method of forming a sheet of paper involves adding the pulp, or "stuff," to a vat of clean, fresh water. The fiber content in the vat is about one or two percent. The sheet of paper is formed by a man or woman known as a "vatman," who holds the mold and deckle above the vat and dips it into the suspension of fibers. With a steady, uninterrupted movement, the vatman brings the mold and deckle up out of the vat in a horizontal position. He or she quickly shakes the mold slightly, from side to side and back to front, as the water drains through the screen surface. The action sends a ripple of pulp in both directions. It helps to even the surface of the pulp and to interweave the settling fibers. The vatman rests the mold on the side of the vat to drain, then removes the deckle and passes the mold to a colleague known as the "coucher." The vatman replaces the deckle on a second mold (a pair with the first) and continues, making another sheet of paper.

COUCHING

The process of transferring a freshly made sheet of paper onto another surface is called "couching." The term comes from the French *coucher*, meaning to lay down. Couching allows multiple sheetforming with a single mold. In Western papermaking, each sheet is couched onto a damp blanket called a "felt." A stack of freshly couched sheets of paper, alternated with felts, is known as a "post." A post usually consisted of six quires (a quire contained 24 sheets). Today, a post contains 50-100 sheets, depending on production requirements.

The coucher takes the first mold from the vatman and props it against a small post, called an "asp," to continue its draining. He or she then inverts the mold, with the freshly made sheet sticking to it, and presses it against a damp felt, transferring the new sheet of paper from the mold onto the felt. The coucher returns the mold to the vat, covers the couched sheet with a second felt, and takes the second mold from the vatman ready for couching.

Nagashi-zuki sheetforming is quite unlike the Western practice. Besides the mixture of fiber and water, a clear mucilaginous extract, called *neri*, is added to the vat. Neri is derived from the roots of various plants, *tororo-aoi* (a variety of hibiscus) being the most common. The use of neri is unique to Japanese papermaking. It gives the water a certain viscosity, which slows the drainage of the vat mixture through the su (see p. 20) during the sheetforming process, and it acts as a formation aid by preventing the long bast fibers from becoming entangled.

The papermaker dips the front edge of the mold, the sugeta (see p. 20), into the vat and scoops up a small amount of the fiber and neri mixture. He quickly tilts the mold so that the solution rushes across the su and off the far side of the mold. He immediately scoops up another "charge" and sloshes the mixture back and forth, and sometimes from side to side, across the su. As the water drains away, he may toss off the excess pulp before dipping the mold again. This process is repeated several times, gradually accumulating layer upon layer until the appropriate thickness is achieved.

To couch the sheet, the papermaker opens the hinged deckle and carefully removes the flexible su. He turns to a

Kathryn Clark forming a sheet of handmade paper at Twinrocker. As the water drains through the surface of the mold, it is dexterously shaken from side to side and from front to back to level the sheet.

The rhythm of the vatman, who makes the sheet, and the coucher, who transfers the freshly made sheet onto a felt, establishes the production of the vat.

table behind him, and having rotated the screen so that the newly made sheet faces the couching surface, he lines up one edge and gently lowers the su across the surface until it lies flat. He then picks up the near edge of the su and carefully peels it away in a continuous movement, leaving the new sheet behind. He returns to the vat with the su and continues, making a second sheet. He couches each new sheet directly on top of the previous one, without an interleaving felt. Sometimes a thread is laid between the sheets – about ¼in. (6mm) from the near edge – which serves as a marker to facilitate separating of each layer after the post has been pressed (see p. 44–7). After a day's papermaking, a nagashi-zuki post can contain 250 to 500 sheets.

CHOOSING AND USING FELTS

Felts are an important part of Western papermaking. They support the paper when it is couched and help to facilitate the removal of water when the sheet is pressed. The texture of the felt will be impressed onto the sheet of paper and can be used to produce a distinctive surface pattern on the finished sheet of paper. Traditional papermaker's felts are 100 percent woven wool and are extremely absorbent. However, they are prohibitively expensive when new. Suitable alternatives are old woolen blankets, cotton diapers, and various nonwoven or synthetic materials, such as interfacing

TECHNIQUE

Forming your sheets

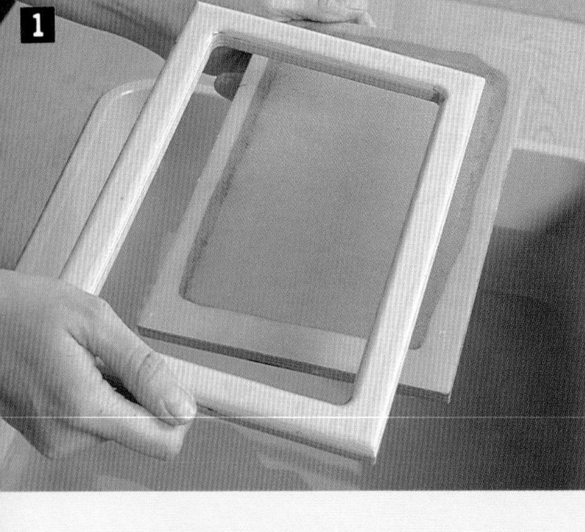

1 The mold surface must be uppermost and the deckle is placed on top.

2 Hold the mold and deckle in a vertical position above the vat, with your hands on the shorter sides.

3 Dip the mold straight into the far side of the vat, bringing the bottom edge toward you.

fabric (the sew-in variety). Thick sheets require heavier weight felts than thinner papers. Interfacing fabric is available in different weights and can be selected accordingly.

Having the correct amount of water in your felts will greatly assist the transfer of the sheet from the surface of the mold. The felts should be well dampened, but not dripping wet. If you find the transfer is incomplete, the felts may not be wet enough. It is a good idea to have two sets of felts, one for white paper and one for dyed pulp. Felts should be carefully washed after use and hung up to dry to prevent mildew.

COUCHING YOUR SHEETS

The process of transferring a newly made sheet of paper onto a dampened cloth, or felt, requires a little practice. Although your first sheets may be difficult to couch on a flat surface, you will find that the process (using a continuous rolling movement) becomes easier as the post of sheets and interleaving felts assumes a natural curve. You can create a slightly rounded surface on which to start couching by folding several couching cloths into a small mound beneath the middle of the first felt. This should be carefully removed before the post is pressed (see p. 45-6).

4 As it reaches a level horizontal position, lift the mold straight up out of the vat in a single continuous movement, without hesitation. As soon as the mold is lifted clear of the vat, and while the water is draining, gently shake the mold from back to front and side to side. If you shake too vigorously, or continue after the water has drained, the pulp will sludge up in the direction of the shake, and the sheet will be unevenly formed. Pulp liquidized in a blender can drain quite quickly. Using a finer mesh surface on the mold will slow the drainage and give you more time to execute a proper shake.

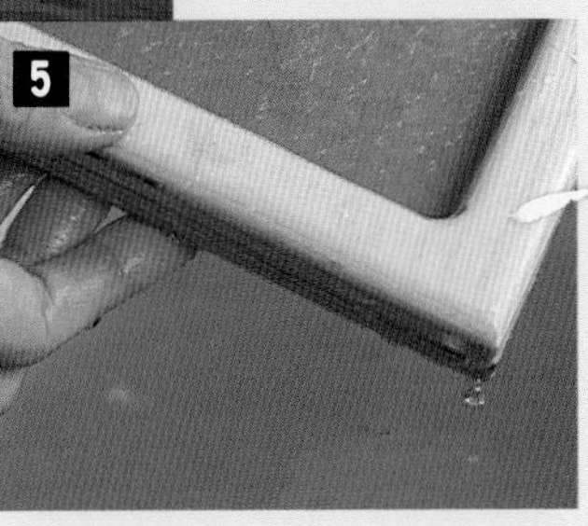

5 Hold the mold above the vat until the water has drained completely.

6 Rest the mold on the side of the vat, or a flat surface beside it, and remove the deckle carefully, making sure that no drops of water fall onto the wet sheet.

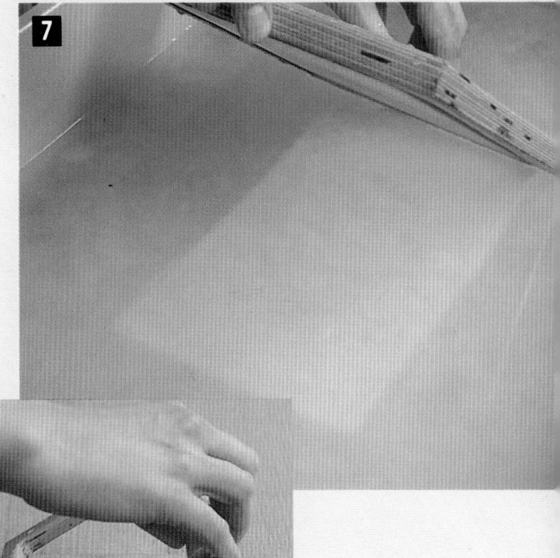

7 If the sheet is unevenly formed or damaged by drops of water from the deckle ("vatman's tears"), turn the mold face down and lightly touch it to the surface of the water in the vat ("kissing off"). Stir the pulp back into the vat, dispersing the fibers evenly.

TECHNIQUE

How to couch

1 Couching should be done on a firm surface. A piece of board, about 4in. (10cm) wider than your mold on all sides and suitably waterproofed, will provide a good surface and can be used as a pressing board when you have finished couching. Your couching cloths (felts) should be about 2in. (5cm) wider than the four sides of your mold. If you are using a light- to medium-weight felt, it is a good idea to put a thicker cloth or several thinner layers underneath it to provide a receptive surface for couching your first sheet.

2 Hold the mold upright at the side of the felt and start couching from the longest edge, using a swift, firm rolling action across the surface of the felt. An uneven pressure, or slight hesitation, can cause air bubbles to become trapped between the felt and the sheet and leave part of the sheet stuck to the mold surface.

3 Place another dampened felt on top of the couched sheet, making sure there are no wrinkles in it. Then make a second sheet of paper.

4 When this sheet has drained and you have removed the deckle, line it up along the same edge and couch it directly on top of the first sheet. Cover the paper with a dampened felt as before and continue layering the wet couched sheets and dampened felts.

5 When you have finished couching, cover the top sheet with an extra felt. Place a second board, the same size as the base board, on top of the pile. If you are using a lightweight couching cloth, cover your last layer with an additional, thicker cloth before you lay down the top board.

After pressing, the damp sheets of paper are separated from the felts by the "layer" and laid on top of one another against a sloping stool.

Pressing and parting

Pressing creates a physical bond between the fibers and starts the drying process.

After couching a quantity of sheets with alternating felts, the post (see p. 46) is taken to a press, where much of the water is expelled under considerable pressure. The function of the press is to remove as much water as possible before drying, and to help bond the fibers into a strong sheet. Traditional wooden screw presses were used for this task until more powerful hydraulic presses took their place.

As soon as the process of de-watering is complete, the post is removed from the press, and a third member of the papermaking team, the "layer," separates the felts and damp sheets. The sheets are laid carefully, one on top of the other, in a "pack" against a sloping stool, and the felts are returned to the coucher for re-use. The sheets are frequently given a second, longer pressing with dry felts, using less pressure. They might then be parted and restacked, or "exchanged" (see p. 49), without felts and pressed again several times. This further pressing diminishes the mark of the felts and smooths the surface of the paper.

Early Eastern pressing methods consisted of a simple lever press hung with weights, or several large stones placed on top of the post of paper on a pressing board. The traditional Japanese way of removing the water from a post of freshly couched sheets of paper is far more gradual than the equivalent Western process. When enough sheets have been couched, the post is covered and left to drain overnight. The following morning, the post is pressed slowly for several hours under light pressure. The pressure is gradually increased in such a way as to remove the excess water without causing the sheets to stick together.

PRESSING AND PARTING PAPER

Pressing a wet sheet of paper can be done by one of several methods. Once the sheet has been transferred from the mold, you can place it between acid-free blotters or absorbent cloths, and sponge out the excess water by hand. Sometimes a rolling pin or mangle is used, but this can distort the wet sheet by forcing the pulp to travel in the same direction as the water. A simple press, where an even surface pressure is applied on top of the sheet, or post of sheets, forces the water from the center out toward the edges and reduces the possibility of displacing the pulp.

A heavy weight – such as a stack of evenly placed bricks, or a centrally placed bucket filled with water or sand

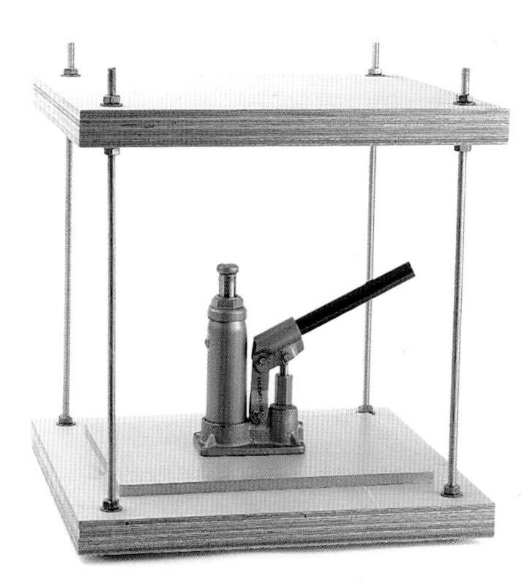

A home-built press (above) can be assembled using a small hydraulic jack, plywood boards, threaded rods with nuts and washers and a small metal plate or piece of wood. Place this between the top of the jack and the bottom surface of the top platform to spread the pressure.

Massive wooden screw presses (left) were used to squeeze out as much water as possible from a post of freshly made sheets of paper.

With a 20-ton hydraulic jack, the Howard Clark press (above) is made of welded structural aluminum and is capable of pressing sheets of paper up to 22×30in. (55×75cm).

– can be positioned on top of your covered post. More weights can be added to increase the pressure gradually and force the water through the post. A substantial amount of water will be removed by pressing. Stand your press on a raised platform in a tray to contain the water as it is squeezed out.

A post can be a small one, with perhaps only 10-15 sheets to begin with. When building up a larger post, the natural curve of the couched sheets becomes more pronounced. If the post is too large, and the curve becomes too steep, the sheets may be distorted during pressing.

MAKING A PRESS

You can make a simple press along the lines of a traditional flower press. Take two pieces of ¾in. (2cm) plywood or blockboard, cut at least 2in. (5cm) larger than your post of felts and paper, and protect them with several coats of polyurethane varnish. To achieve the necessary pressure, you can use two evenly spaced C-clamps on each side of the boards, or you can drill a hole in each corner of the plywood to accommodate a wing-nut screw or galvanized carriage bolt. Alternatively, you could make battens to place across the boards. Use four equal pieces of 2×1in. (5×2.5cm) pine-

A newly formed sheet of paper contains about 90 percent water. Pressing speeds the drying time and improves the quality of the sheet. There are several methods of pressing. You can place your paper between acid-free blotters, or absorbent cloths, and sponge by hand (left).

You can use a rolling pin to press the paper, but this can distort the wet sheet (below).

You can place a heavy weight on top of the covered post and gradually increase the weight (below).

wood, 4in. (10cm) longer than the boards. Drill a hole in the middle of each end to take carriage bolts with nuts and washers that will tighten the press to the maximum pressure.

Greater pressure still can be obtained by using a traditional screw-operated book press. Or you could build your own mechanical press with the aid of a small hydraulic jack of the kind used for changing car tires. Both methods will speed up pressing and drying time considerably.

USING YOUR PRESS

Leave the post in the press for about half an hour. The bottom sheets and edges of the paper may still be wet if the post has not been pressed for long enough. Remove the post from the press and place it on a dry working surface. Lift the top felt or couching cloth off slowly. Pick up each damp pressed sheet by carefully lifting the corners of the closest long edge, or by lifting two opposite corners toward each other. If you find this too difficult at first, try lifting the supporting felt with the sheet still in place, taking care not to lift the sheet below.

You can transfer your post of paper and felts to a simple press.

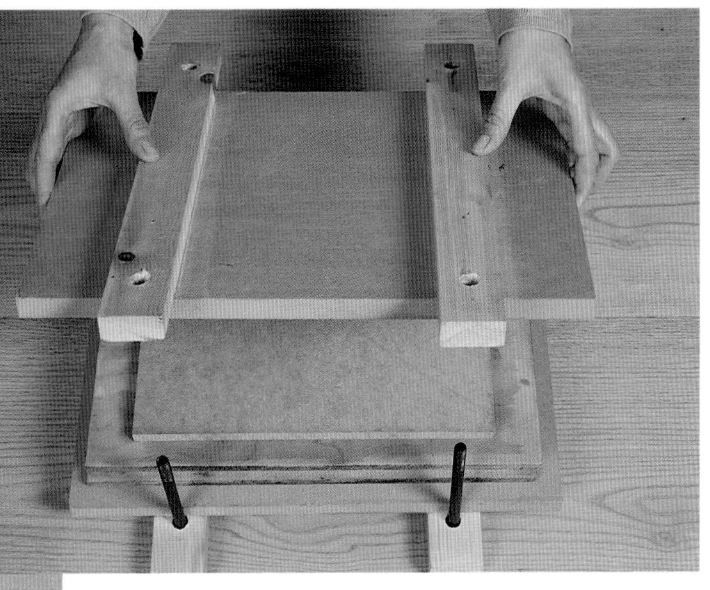

After removing the post from the press, carefully separate each sheet from the one below by lifting diagonally opposite corners (below).

Screw down the bolts (above). After about 10 minutes, give the screws a few extra turns to increase the pressure. Ideally, pressing should be for about 20-30 minutes with gradually increasing pressure, or until all the excess water has drained through the felts.

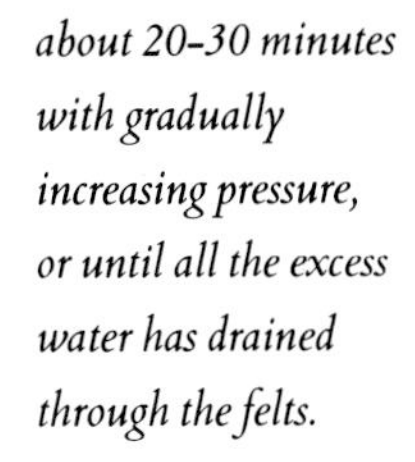

In China, bamboo paper is left to dry on the ground after pressing, and it is not unusual to see whole papermaking districts literally covered with drying paper.

Drying the paper

After pressing, the Japanese papermaker brushes each one onto a smooth board to dry in the sun, or onto a heated metal sheet to dry indoors.

After pressing, the still-damp sheets were taken to an upper story – the drying loft – and hung over the horse- or cow-hair ropes coated with beeswax that stretched across it. Drying lofts were carefully ventilated by means of wooden louvers, which could be adjusted according to the humidity inside the room and weather conditions outside.

Successful drying is best achieved slowly. Undesirable shrinkage and warping, or cockling, at the edges of the sheets can occur as the fibers themselves shrink laterally during drying. Hanging the damp sheets in "spurs" – several sheets lightly pressed together into a small pack – helps to lessen the distortion of each individual sheet as it dries. When the paper is completely dry – it may take a week or longer – the sheets are easily separated from the spur and can be stacked and lightly pressed to reduce any distortion. In the past, a ridge of paper, known as a "back mark," was often caused by the hair rope over which the paper was hung.

Various heating systems were introduced during the 18th century to speed up the process of loft drying. Modern papermaking machines have a "dry end," consisting of drying cylinders and a set of rollers. The paper is passed, dried, and smoothed between these, thereby solving the problem of drying paper quickly while maintaining its flatness.

Oriental papers are dried in various ways. The Chinese leave the sheets outside on the ground, and in Nepal they are left to dry on the mold, whereas in India damp sheets are sometimes brushed onto plaster walls. Japanese papermakers traditionally separate damp sheets from the post after pressing and brush them onto seasoned wooden boards to dry outside in the sun. This often results in a subtle lightening of the paper. It is more usual today, however, for the damp sheets to be brushed onto a heated metal drying surface.

DRYING YOUR PAPER

Drying methods have a considerable effect on the surface and handling qualities of the finished paper. Sheets that are pressed and then dried on boards will be flat and smooth with a slightly different texture on each side. Sheets that are left to dry on the mold without pressing will also have two characteristic surface textures: the side which dries against the mold will be smoother than the outer side. These and similar methods, which resemble the Oriental model, are known as restraint drying.

Drying lofts had ropes like washing lines, over which the paper was hung to dry. After drying the sheets are stacked under light pressure and "exchanged" regularly so that the air gets to the surfaces of all sheets.

RESTRAINT DRYING

If you have separated your post by lifting each felt and damp sheet together, you can leave the paper to dry on the felt – a simple form of restraint drying. Drying the sheet on the felt is, however, quite time-consuming. A quicker method is to place the felt, sheet side down, on a smooth, clean, flat surface. Sheet glass, Formica, plexiglass, or wooden board are all suitable, but it is important to make sure that wood has been sealed or is well seasoned to prevent it from staining the damp paper. Roll the back of the felt gently but firmly with a small roller or rolling pin. Lift the felt carefully from the back of the sheet, leaving the sheet stuck to the drying surface. The paper should remain flat on the surface, without lifting or curling off, until it feels completely dry.

Lift a corner of the paper and peel it away from the board. If part of the sheet begins to pull away before the rest is dry, use a plant sprayer filled with fresh water to spray a fine mist over the entire sheet. Then reapply the sheet to the drying surface. Once removed, dry sheets can be stacked between boards and stored under a weight such as a covered brick or a heavy book.

EXCHANGE METHOD OF DRYING

If you have separated your damp pressed sheets from their supporting felts, transfer them onto dry felts, making sure there are no creases in the paper, and return them to the press for a slow pressing overnight. When you remove the sheets from the second pressing, stack them one on top of the other without interleaving felts. Place a single dry felt at the bottom and top of the stack and press it again, more lightly, between the pressing boards. You can repeat this pack pressing several times, separating and restacking the papers each time, until the sheets are dry and the surface becomes quite smooth.

Alternatively, after the second dry felt pressing, the still damp sheets can be interleaved with acid-free blotters, cut to the same size as the sheets. Keep the post under a light, even pressure, and exchange the blotters regularly for dry ones as they draw the moisture from the sheets. Once the sheets are completely dry, place them under a light weight and store them in a cool, dry room.

A DRYING BOX

As a variation on the blotter method, you can make your own cabinet-style plywood drying box, which uses the same sequence of layering as a traditional flower press. The box should be left open at the front and have a removable top. Interleave the damp paper with blotters, but include additional packing material, such as a double-wall corrugated board, porous foam, or stiff wire screen, at regular intervals as "spacers" between the blotters. Layers consist of spacer board, blotter, paper, blotter, paper, blotter, paper, blotter, spacer board, and so on – depending on the dampness of the sheets. Packing materials for drying sheets or low relief artwork in a paper dryer are available from papermaking suppliers.

Stack the post in the drying box, leaving a narrow gap as an air space between the back wall and the drying stack. Mount a small fan flush with the inside back wall of the box. This creates a current of air which draws the moisture out of the stack and speeds up drying time. The stack can be kept absolutely flat, either by the weight of the top board or by means of winch straps on opposite sides of the stack.

LOFT DRYING

Sheets which are not restrained during the drying process are more likely to warp or shrink. On the other hand, the fibers have a greater opportunity to bond together as the sheet dries. This results in a strong sheet of paper with good folding strength. You can loft-dry your sheets after pressing by hanging them over a wooden clothes dryer or length of plastic washing line.

DRYING TIME

Whichever method you use, drying times vary according to the weight, thickness, and size of the sheets, as well as the surrounding temperature and humidity. Paper should always be dried slowly. Use a fan to keep the air circulating around the sheets rather than an artificial source of heat directly on them.

Having separated your sheet with its supporting felt from the "post" (left), lay it carefully on the drying surface (sheet side down) and roll the back of the felt (below).

Lift the felt from the back of the sheet, which will have adhered to the drying surface (above). Peel the sheet off the board when it has dried completely (right).

A mechanical method of imparting a smoother surface to paper by means of a glazing hammer soon replaced earlier polishing by hand.

Sizing and finishing

The names "rough", "not" and "hot-pressed" are used to describe the traditional finishes of handmade paper.

Until paper has been sized, it is known as "waterleaf" paper. It is generally very absorbent if wetted again. Sizing makes waterleaf paper less permeable and is applied according to the end use of the paper. It reduces the bleeding or feathering of any water- or oil-based media applied to the paper's surface. It can also improve the sheet's smoothness and protect the fibers from dirt and pollution.

Traditionally size is applied to the sheet after it has been pressed and dried. This is called external, tub, or surface sizing. Common surface sizes are vegetable starches made from rice and wheat, and gelatin derived from animal tissue or bones. Nowadays, size is frequently added before the sheet is formed, either during beating or in the vat. This is variously termed internal, beater/engine, or stock sizing. Two of the most commonly used internal sizing agents were alum and rosin. However, their high level of acidity limits the life of paper, and a neutral pH liquid size (alkylketene dimer) is now more widely used.

Finishing is a term for the final processes that affect the surface texture of the dried paper, increasing its smoothness or gloss, for example.

Once dried, European handmade sheets were taken down from the ropes in the drying loft and loosened from one another if they had been dried in spurs, ready for sizing. Gelatin was the traditional ingredient of size, having replaced the starch size inherited from Arab papermaking. The size was prepared by boiling the assorted remains of slaughtered animals. The resulting gelatin was strained and diluted before use. Gelatin size (increasingly mixed with alum) was warmed in a sizing vat, and spurs of paper – about 50 sheets – were immersed in it. The spurs were assembled into a post of size-soaked paper and pressed to force the size through the entire post. Out of the press, the spurs were separated into single sheets and loft-dried as before. After this second drying, the individual sheets were stacked and pressed in a dry press, then restacked and pressed again before proceeding to the finishing treatment.

Hand methods of surface sizing and finishing continued in the West until the introduction in the 16th century of the glazing hammer, and then later in the 18th century, mechanically-driven glazing rolls, which imparted a glazed finish to the paper as it passed between the rollers.

FINISHES

Handmade paper can have three basic finishes: "rough," the surface texture of a sheet which has been pressed once between felts and dried without any further smoothing or pressing; "not," or "not hot pressed," the result of separating the sheets from the felts and pressing them again without interleaving felts; and "hot pressed," or "HP," a glazed surface that comes from passing the paper between hot metal plates or rollers.

Traditionally Japanese papers are not usually sized. When additional surface resistance is required for a particular printing technique, an animal hide glue and alum size is brushed onto the paper after it has dried. For special paper, a camellia leaf is sometimes used to shine the top side of the paper as it lies on the drying board.

SIZING AND FINISHING PAPER

The sizing method you choose will depend on how you intend to use your paper. Paper for printmaking, watercolor paper, and calligraphy paper all require different amounts of sizing. A strong sizing application results in a harder, less absorbent paper. A light sizing produces a paper which will show more feathering or "bleedthrough" of such media as paint and ink.

INTERNAL OR EXTERNAL SIZING

First you need to decide whether you want to size your paper internally — by adding a sizing agent to your pulp — or surface-size it after it has been pressed and dried. In either case, there are several sizing applications to choose from. Neutral pH alkylketene dimer emulsion for internal sizing is available in a concentrated form from most papermaking suppliers. It simply needs to be diluted with water and added to the pulp during the last few minutes of the beating or mixing cycle.

Contemporary papermakers have experimented with a variety of internal and surface sizes, including gelatin, household and vegetable starch, methylcellulose, and acrylic and silicone sprays.

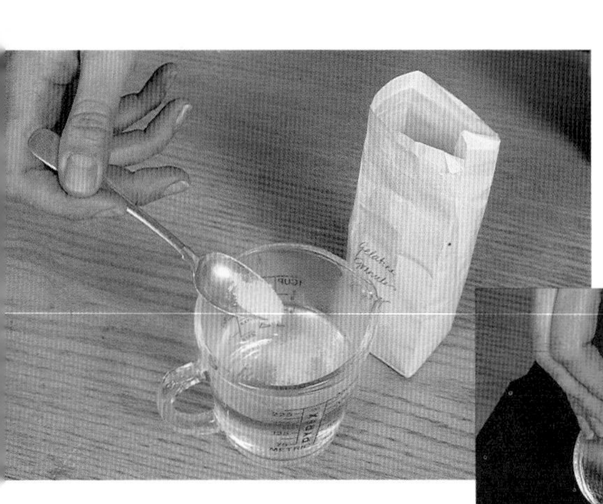

Dissolve a measured amount of gelatin in hot water.

Pour the solution into a shallow tray.

Immerse the sheets of paper in the size.

Some substances, such as alum and rosin size, cause the paper to discolor and are best avoided. Rice or wheat paste and gelatin sizing can make paper prone to insect or fungal attack in humid conditions. Others stiffen the paper by "gluing" the fibers together, but do not increase the surface stability for writing.

Surface sizes are applied to dry "waterleaf" paper. If you can, leave your sheets to mature for several weeks before sizing them. The dimensional stability of the sheet will be increased, and the paper will be less likely to disintegrate in the sizing solution. The size can be brushed or sprayed onto the paper, or the sheets can be tub-sized in a shallow vat.

USING GELATIN

Gelatin sizing can greatly improve the strength and flexibility of a short-fiber sheet. Use a good quality gelatin (available from photographic or fine art suppliers). Mix a 2-3 percent solution by adding ¾-1oz. (20-30g) of gelatin to a quart (liter) of hot water and stir until the gelatin has completely dissolved.

Pour the solution into a shallow plastic or enamel tray, and carefully immerse the paper. The sheets can be immersed one at a time or in a larger stack, depending on the thickness of the sheets and the depth of the tray. There must be enough size in the tray to cover the sheets completely. Use a piece of wooden doweling to lift the soaking sheets and draw them out of the tray. Lay the sheets, one on top of the other, on a felt-covered pressing board. Place a second felt and pressing board on top of the stack and press gently to force the size through the sheets. Stand the press in a tray to catch the excess size, which can be used again. Then strip the sheets carefully from the pack – remember to wash the pressing felts thoroughly and hang them to dry. Restack the sheets so that the middle ones are on the outside. Do not place the stack under any pressure other than the weight of the paper itself. Repeat this process, changing the position of the sheets each time, until some of the moisture has left them and they become easier to handle. Then either hang the sheets up, or place them on a clean, flat surface to dry. When they are completely dry, restack them under light pressure and store them in a cool space.

If you are sizing a number of different sheets, perhaps of varying thickness, it is best to soak them one at a time. If you want to size several batches, or larger packs that need to soak for a few minutes to absorb the size, place the sizing tray in a bath of hot water (kept at about 122°F or 50°C) to prevent the gelatin from cooling and setting during the sizing operation. It is a good idea to place a sheet of plexiglass (slightly larger than your paper) on the bottom of the tray beneath the sheets to help you lift out a larger stack.

Use a piece of wooden doweling to lift the sheets out of the tray.

Lay them carefully on a felt-covered pressing board.

Separate the sheets after they have been lightly pressed.

CHAPTER

3

Papermaking Variations

Hugh O'Donnel: Weapons of Desire, Bedford Series III. Colored and pressed paper pulp, (left). Vito Capone: Carta scolpita. Cotton rag paper "engraved" with a sharp tool, (right).

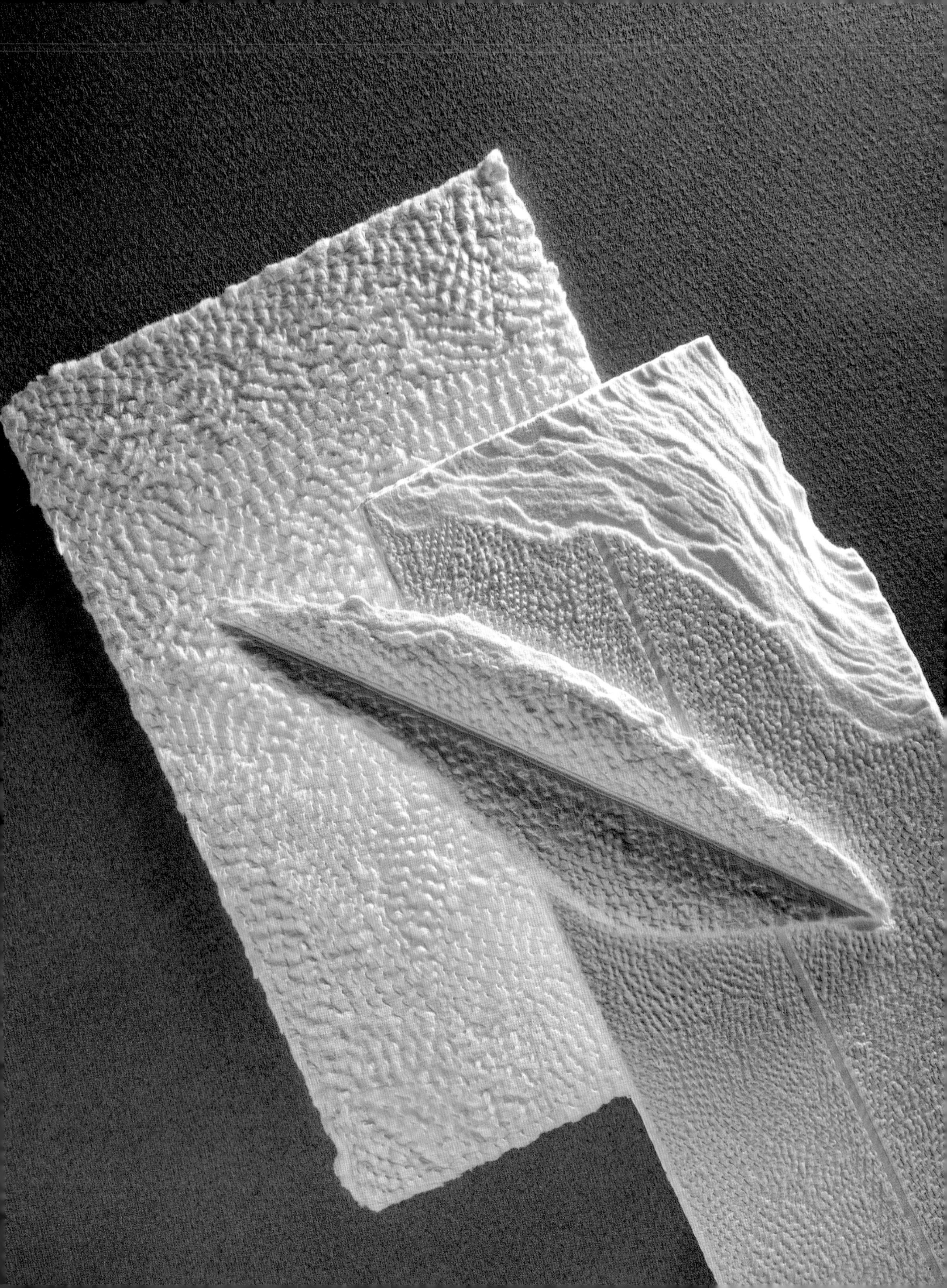

Kathryn Clark: Constructing Cube in Space. Handmade paper pulp. Lamination with embedments. 24×38in. (61×97cm).

Laminating

Peter Sowiski: Ubee, Saturation, Stealthy (installation). Colored pulps built up in shaped and overlaid laminations. 60×92in. (152×233cm), (above).

Laminating is a simple technique of couching one or more newly formed sheets of paper on top of each other. During pressing and drying, the fibers of each layer bond together, creating a single sheet. Japanese handmade papers are formed by gradually laminating a thin sheet of paper on the mold surface. You can take advantage of both techniques to create subtle and interesting overlaid effects.

Laminating (also called multiple couching) can be used in a variety of ways. For example, two layers of different colors produce a double-sided sheet, or several overlapping layers can create a soft, misty effect that is similar to that achieved using superimposed washes of watercolor. However, laminating sheets made from different fiber pulps can cause problems during drying, particularly if one is a long fiber and another is much shorter, or if one fiber has a tendency to shrink more than another. To prevent distortion, it is best to board-dry these laminated sheets.

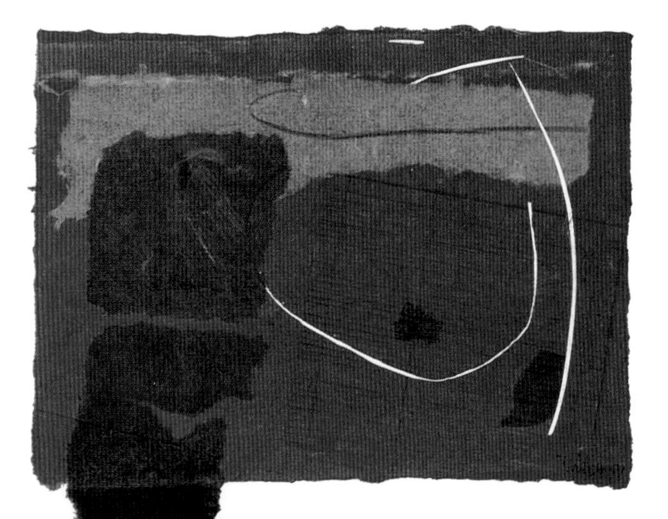

Therese Weber: Spring Leaves. Handmade paper (cotton and abaca) lamination with embedments. Colored with pigments. Pressed and airdryed. 12×8½in. (30×22cm).

TECHNIQUE

Different Effects

1 Make a piece of paper. Couch it onto dampened felt on your couching board. Devise a registration line to correspond with the nearest edge of your mold as it is positioned on the couching felt. This could be a piece of thread with the ends secured under the couching board.

2 Make a second sheet of paper. Couch it exactly on top of the base sheet, making sure that you line up the edge of the mold against your registration mark. When pressed, this will give a thicker sheet of the same color.

3 Without a method of registration, you are unlikely to achieve an exact multiple couch.

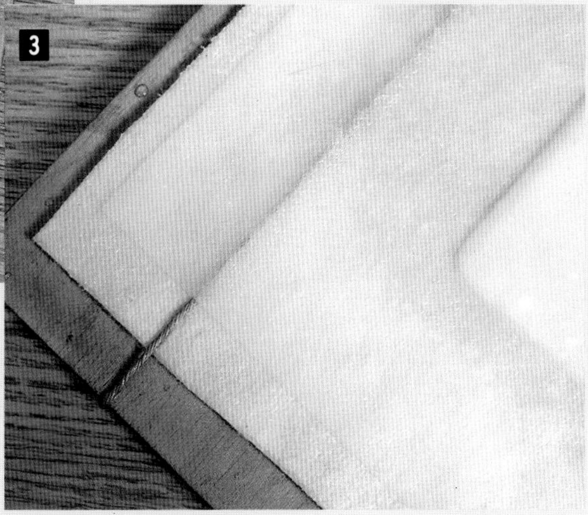

1 To give a double-sided sheet when pressed, make a base sheet and couch it. Make a second sheet using a different-colored pulp, and couch it on top of the base sheet.

2 Couching additional layers will create a subtle intensity of color in the laminated sheet.

YOU WILL NEED

- *prepared pulp: 2–3 colors*
- *mold and deckle*
- *couching cloths*
- *pressing boards*
- *pencil or short length of thread (for registration line)*

TECHNIQUES

- *sheetforming*
- *couching*
- *pressing*

1 Another technique involves altering the position of second or subsequent sheets, say by couching the edge of one sheet along the edge of another to extend the dimensions of the sheet.

2 You can also change the angle of each sheet as it is couched onto the one below to create a non-rectangular piece of paper.

1 By removing the deckle from the mold, you can build an image within your sheet by laminating washes of pulp and couching off-register. Set the deckle aside and dip your mold partially into the pulp at a 45° angle, picking up a thin wash of pulp.

2 Lift the mold straight out at the same angle. Couch the pulp onto a base sheet strong enough to support subsequent laminations.

3 You can use both edges of the mold to pick up two washes of the same, or a different, colored pulp.

4 Continue to build up overlapping layers.

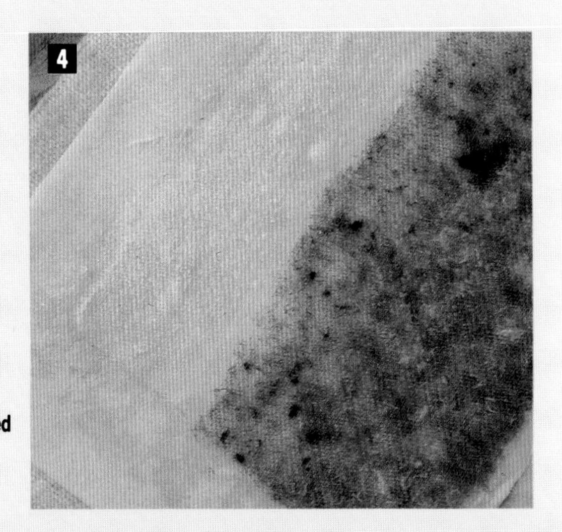

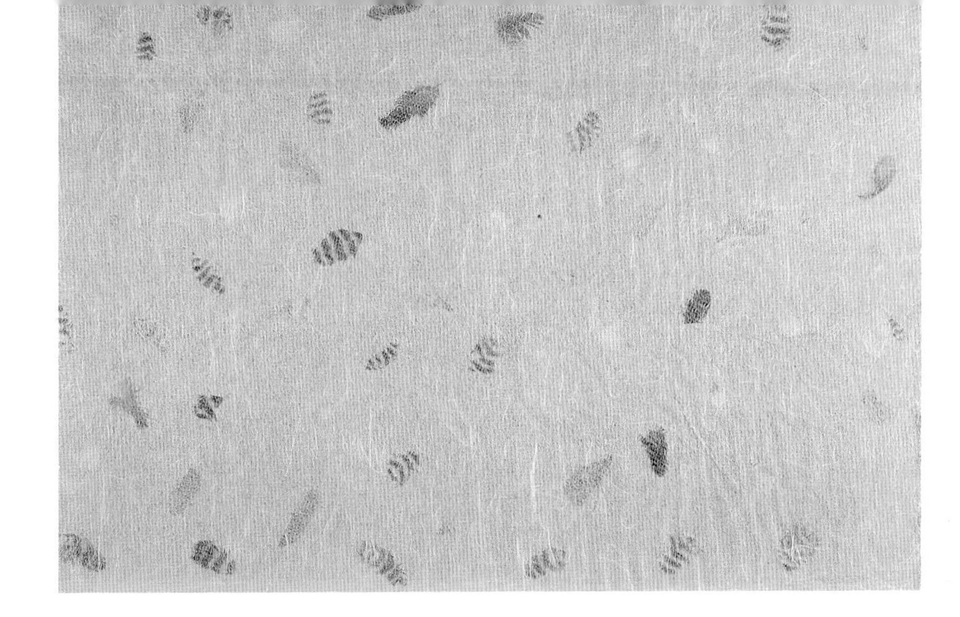

Japanese "rakusui" paper with feathers embedded between the layers, (detail).

Embedding

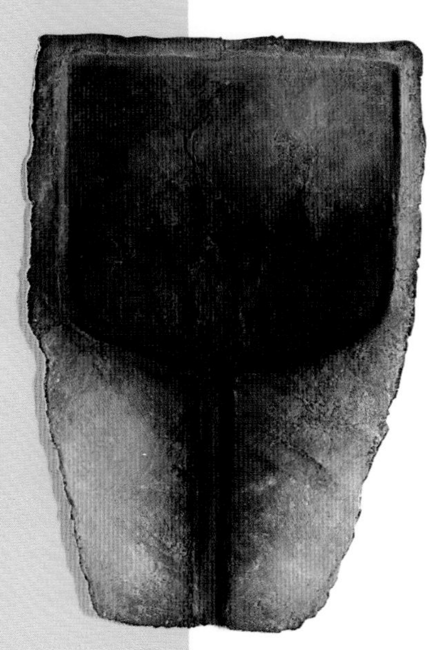

Carol Farrow: Evier 1. Damp paper laid over a spade and allowed to dry. Molded sheet removed and painted with acrylic wash. 47×32in. (119×81cm).

Embedding is a process of incorporating materials such as threads, leaves, twigs, feathers, and printed images into couched sheets of paper. Embedded elements are held in place by the fibers and the bond created during pressing and drying rather than by an adhesive.

You can embed elements in your paper while it is still wet on the mold, or, using a slightly different technique, you can couch a base sheet and place the material to be embedded on it, then couch a second sheet over the applied elements. Part of the top sheet can be torn away, partially revealing the embedded elements; they can either project from the edges of the sheet or, using an opaque pulp, they can be completely concealed between the laminated sheets.

Alternatively you could couch a translucent layer, or a very thin wash of contrasting pulp over the inclusions, so that they are more clearly visible. You could also try dribbling a small amount of pulp over the inclusions to create a bond with the base sheet when the piece is pressed. Paper with embedded elements should be dried on a board after pressing. This will prevent the paper from wrinkling around the inclusions and distorting the sheet.

Some papers deteriorate with age and, if you are concerned about embedded papers which may be acidic, you should consider neutralizing them. There are several methods and formulations of solutions available. Also, it is advisable to check the color fastness of any embedded papers before using them. The dye from colored tissue paper, for example, runs excessively.

TECHNIQUE

Revealing an Image

1 **Tear a colored photocopied image – here a picture of a cardoon plant – into sections. (Photo-copies can be made on acid-free paper.)**

2 **Lay the sections on a base sheet made from the fibers of the cardoon plant.**

3 **Place strips of masking tape across the mold to correspond with the position of the plant image on the base sheet.**

4 **Make a second sheet of paper. The pulp will only form on the untaped areas of the mold where the water can drain through the screen.**

5 **Couch these strips of cardoon paper on top of the plant image.**

6 **When pressed, the base sheet and top strips will form a bond where they meet and hold the photocopy.**

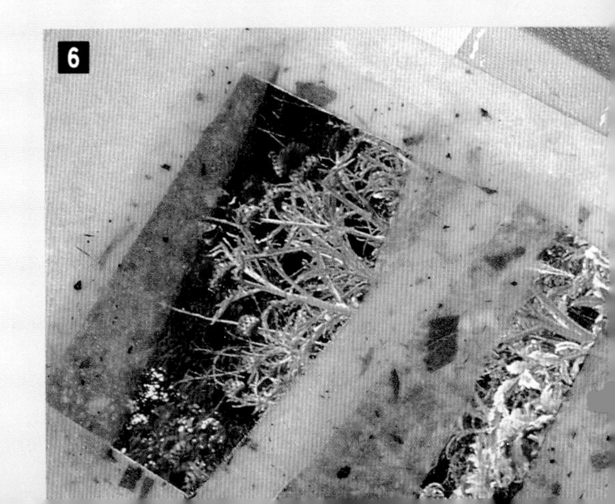

- *prepared pulp*
- *mold and deckle*
- *couching cloths*
- *pressing boards*
- *thin materials to place between layers*
- *masking tape*

TECHNIQUES

- *sheetforming*
- *couching*
- *laminating*
- *pressing*

TECHNIQUE

Embedding an Image

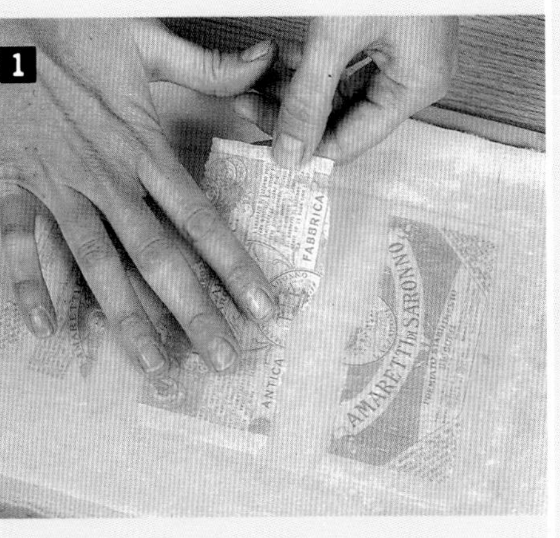

1 Position torn pieces of paper, such as this printed tissue wrapper, on a base sheet.

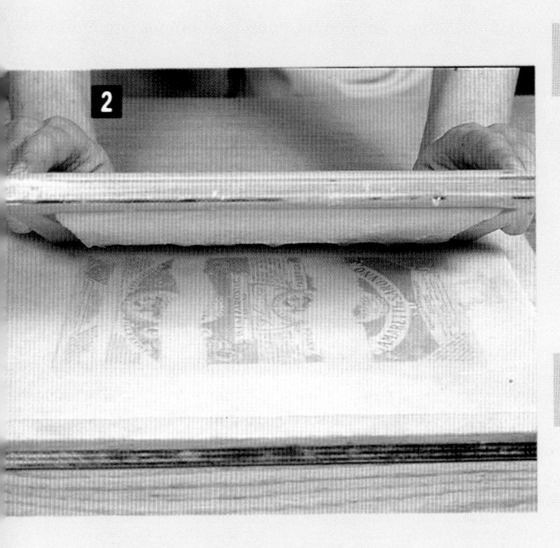

2 Couch a thinner top sheet over the fragments.

3 The embedded image becomes an integral part of the laminated sheet.

TECHNIQUE

Colored Strips

1 Couch strips of a contrasting pulp onto a base sheet and stretch lengths of thread on top of them.

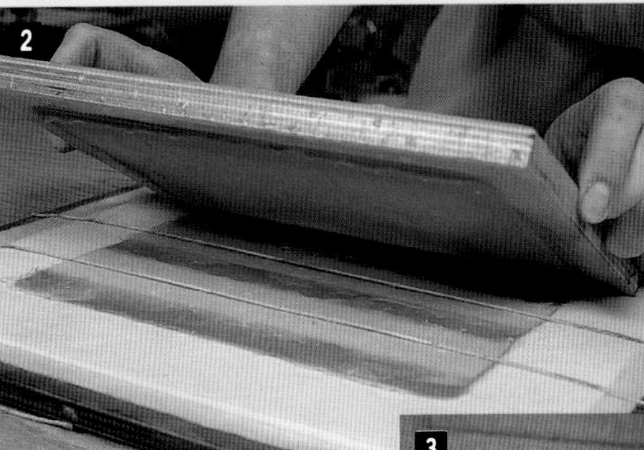

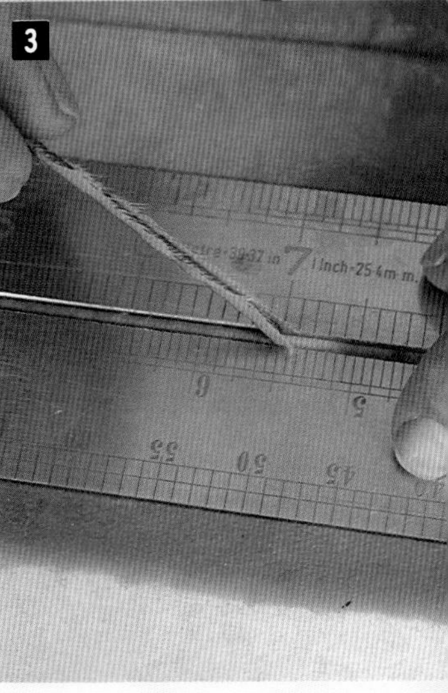

2 Couch a top sheet - the same or a different color from the base sheet - over the threads. Press.

3 Using two metal rulers, tear the threads out of the damp paper to reveal the colored stripes.

Many different types of plain and colored papers, some with long swirling fibers and bark flecks, and others with delicate patterns and soft tones, are used for decorative and artistic purposes in Japan.

Flowers and Foliage

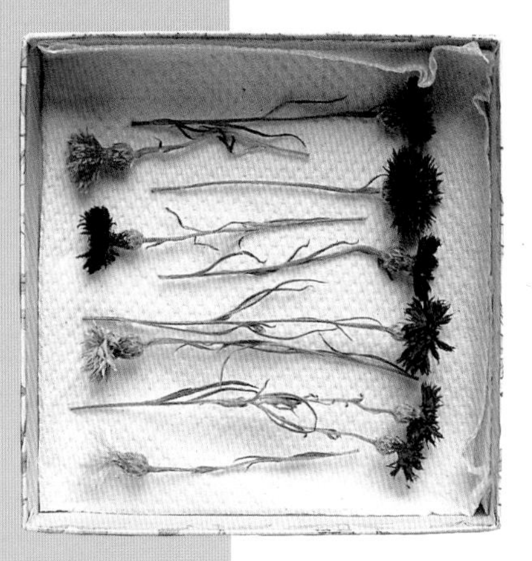

Batchelor buttons and anemones dry well in a microwave. Check after three minutes. If they do not feel and sound paper-crisp, give them a little more time. Leave to cool for a few minutes before removing from oven.

Wherever you live, you can find a wide variety of flowers and leaves, or seeds from trees, ferns, and grasses for making paper with plant inclusions. Sources are not only plentiful but inexpensive: suitable material can often be gathered from lawn cuttings and hedges, or among the sweepings from florists' shops and plant nurseries. Collect only as much as you need: adding just a small amount of flower material or foliage to recycled pulp, or a base stock of another plant fiber, will give your paper a distinctive character.

DRIED FLOWERS

Dried flowers can make good inclusions, especially if they have small petals that will distribute easily in the vat. Pick the flowers when they are in bloom and tie them in bunches immediately. If the flowers are at all wet, allow them to dry out naturally before putting them in a warm, dark, dry place.

There are two important criteria for drying flowers: heat and darkness. If you have a warm cupboard, you could hang the flowers upside down from a coat-hanger to dry there. However, the best results for some flowers come from using a microwave.

You can use professionally dried flowers, but they sometimes contain glycerin and alum, which may cause your paper to discolor. You can reduce the glycerin and alum content by boiling the flowers for 10-15 minutes and then soaking them for several hours. Change the water regularly, rinsing the flowers each time. Rinse thoroughly before use.

PRESSED FLOWERS

Flowers and foliage to be pressed should be picked when the weather is dry. The ▶

PROJECT

Fresh Flowers and Foliage

1 Divide larger foliage into smaller sections, and place them in a saucepan of clean water. Bring to a boil and simmer for 10 minutes. Strain and rinse the leaves carefully. You can test the foliage by placing it between two pieces of clean blotting paper and pressing it for half an hour. If the blotting paper is stained, re-soak the leaves for half an hour and rinse them again.

2 The cypress leaf on the left has lost much of its original color after being boiled and can now be added to the vat.

3 Soaking flowers and foliage will hydrate them so that they will be less inclined to float on the surface of the vat. Unsaturated petals and leaves can be sprinkled into the vat and will be more fully exposed on the surface of the paper. They will be taken up onto the mold during sheet formation and embedded in the wet paper when it is couched and pressed.

YOU WILL NEED

- *fresh or dried plant inclusions*
- *flower press*
- *saucepan*
- *blotting paper*
- *microwave oven (optional)*
- *prepared pulp*
- *mold and deckle*
- *couching cloths*
- *pressing boards*

TECHNIQUES

- *sheetforming*
- *couching*
- *pressing*

4 A short- or medium-length fiber pulp will reveal the inclusions more clearly than a long-fiber pulp, which tends to tangle around the smaller plant parts. A good base pulp to use is an equal combination of abaca and cotton linter. Use a translucent bast fiber if you want the flowers or foliage to be more visible.

◀ best time is in the early afternoon when any dew and hidden water droplets will have evaporated. Be careful to select undamaged, fresh material and press it immediately. A flower press will give the best results, although you can start by using a heavy book or pile of magazines. The most obvious effect of pressing flowers and leaves is that they become two-dimensional. This often reveals unusual aspects of the plant: subtle shapes and colors become more apparent; many colors become richer and darker, and some flowers change their color completely.

FRESH FLOWERS

Most flowers and foliage will release their own pigments when immersed in water, some more readily than others.

Preparing flower petals at the Moulin Richard de Bas. Small petals, ferns, and local grasses are added to the pulp to create a decorative paper that is used for printing poetry, as the endpapers of books, and for making lampshades.

Fresh foliage should be boiled gently for 10 minutes to remove fugitive pigments and any extraneous materials that might leach into or stain the paper. Fresh flowers are generally too delicate to boil and are best pressed or dried.

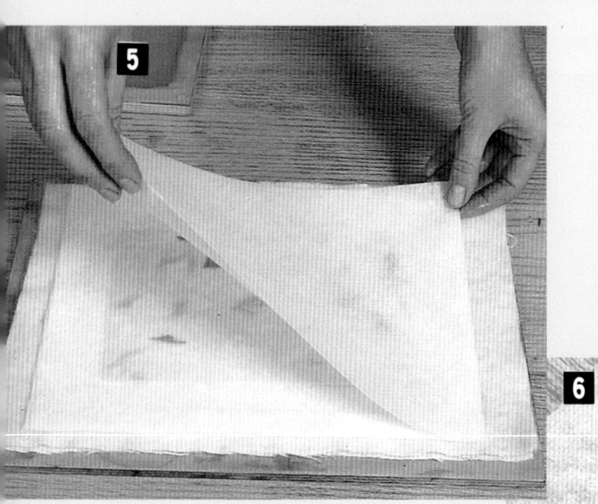

5 Couch the sheets of flower paper onto pellon interfacing, press them lightly, and hang them up to air-dry.

6 The combination of different leaves can create a simple yet eye-catching image. Here, the spray foliage of a Leyland cypress is set off against a miniature autumn rose leaf.

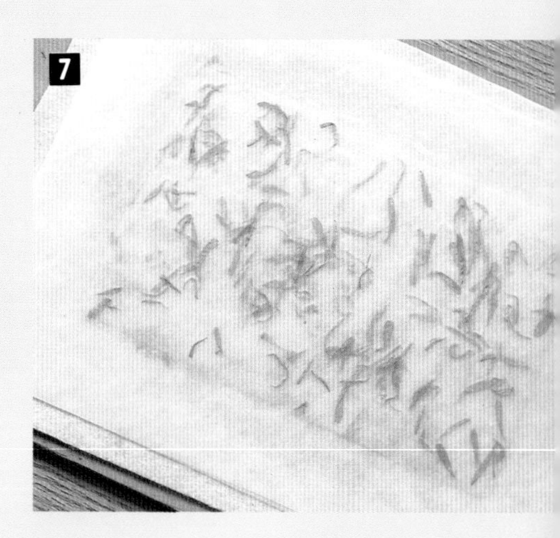

7 The buoyancy of unsaturated flowers can be used to create a dense overall surface effect. Marigold blooms have large quantities of tiny petals which can be picked off the flower-head and distributed in the vat.

Specialty papers (top to bottom): 100% cotton sheet, beater dyed black, mottled with white pulp; Japanese sheet with flecks of gold leaf; decorative paper with embedments; three sheets with added colored fibers.

Decorative Effects

Decorative and printed pattern papers are best known for their role in book design and bookbinding. They can also serve as wallpapers or as linings for boxes, drawers, and cupboards. Fancy work in paper was especially popular in Europe around 1800, when there was a craze for making small decorative objects ornamented with paper and pasteboard.

Western decorative papers include woodblock printed papers, stenciled papers, gilt papers, and marbled and paste papers. All of these are the product of surface techniques — the pattern is applied to the surface of the paper after it has been made. Some Japanese decorative papers, such as *shibori-zome gami* (tie-dyed paper) and *orizome gami* (folded dyed paper), parallel Western surface techniques, but others are created by pulp manipulations. *Unryu-shi* ("cloud-dragon paper") is made by adding long, swirling fibers to the pulp during sheet-making.

With a little experimentation, it is ▶

Japanese "lace" paper (above) comes in many patterns. Nepalese and Bhutanese papers (right) are made by pouring pulp onto a floating mold.

PROJECT

Silk Threads

1 Cut some colored silk into small pieces. Shred some of the pieces into individual threads, by hand or in a liquidizer.

2 Add the fragments to a vat of recycled cotton rag pulp. They create a random and quite spontaneous pattern as the sheet forms.

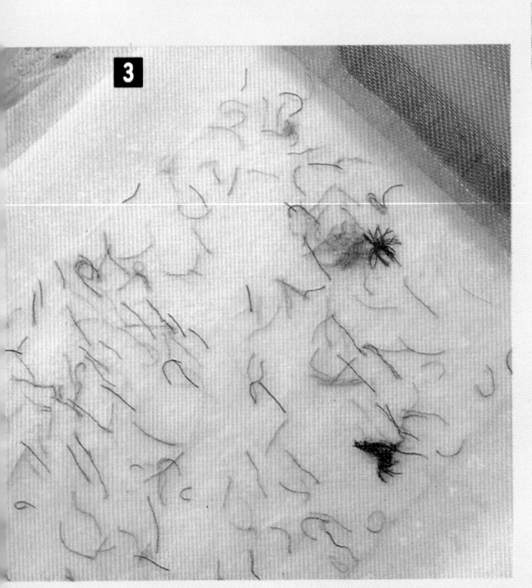

3 Some of the fragments are embedded in the sheet, while others are lifted onto the surface.

PROJECT

Speckled Paper

1 Cut two or more different papers into small pieces and soak them for several hours.

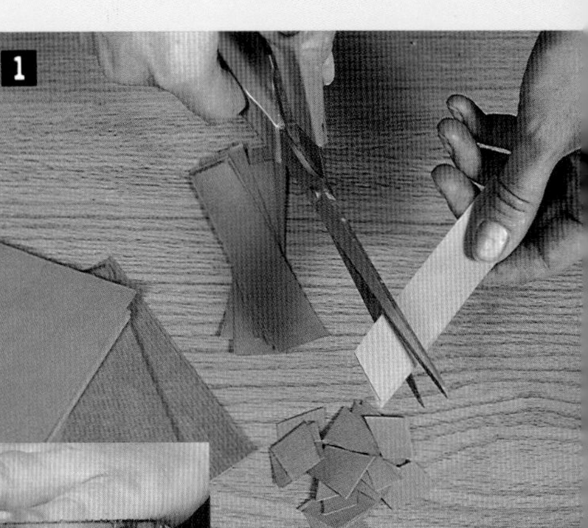

2 Recycle the papers separately in a liquidizer operated in 10-second bursts, leaving some of the paper in identifiable pieces. Then use slightly longer bursts to reduce some of the paper to a finer pulp.

3 Mix the recycled pieces of paper into a plain base pulp and form your sheet. As the freshly couched sheet shows, the different size pieces of colored paper create a pronounced speckled effect.

PROJECT

Lace Paper

1 **Allow a freshly formed sheet of paper to drain for a few minutes before placing it in an upright position.**

"Lace," or "rain," paper can be made by spraying a fine jet of water at a freshly formed sheet of paper while it is still on the mold. If you cover the sheet with a stencil before you spray it, the jet of water will wash out the unprotected areas of pulp to create a pattern of holes.

2 **Fill a plant spray bottle with water and spray the sheet. The water will displace the pulp and make holes in the sheet. You can adjust the nozzle setting to produce a diffuse spray or fine jet of water.**

3 **Couch the sprayed sheet on a contrasting base sheet.**

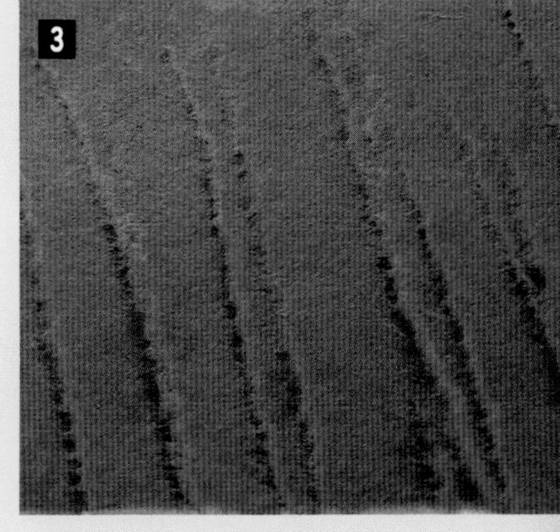

YOU WILL NEED

- *prepared pulp*
- *mold and deckle*
- *couching cloths*
- *pressing boards*
- *colored threads (Silk Thread Paper) or colored papers for recycling (Speckled Paper) or plant spray (Lace or "rain" Paper)*

TECHNIQUES

- *sheetforming*
- *couching*
- *pressing*

◀ possible to create a variety of special decorative effects, using a combination of simple papermaking techniques.

FLOATING MOLD EFFECTS

Nepalese hand papermaking is based on the traditional Chinese method of pouring pulp into a mold that is partially submerged in water. This technique of floating the mold, rather than dipping it, does not require a deckle. The fabric or screen is stretched underneath the frame, which itself acts as a deckle. The sheet is formed in the well of the mold and left to dry outdoors on the screen. When the sheet is completely dry, it is peeled off, and the mold can be used again. Many molds are needed to produce a quantity of sheets in this way.

Sheets made by the floating mold method have a distinctive cloud-like fiber formation, and the technique can be adapted for a wide range of special effects using different colors. You can use an ordinary mold, simply turning it upside down, or make a mold with a slightly deeper frame especially for this technique.

PROJECT

Floating Mold Paper

1 Half fill your vat with water. Place your mold with the screen side face down, so that the water comes through and partially fills the mold.

2 Keep the screen submerged, and pour a measure of diluted pulp into it. Stir the pulp to distribute the fibers evenly.

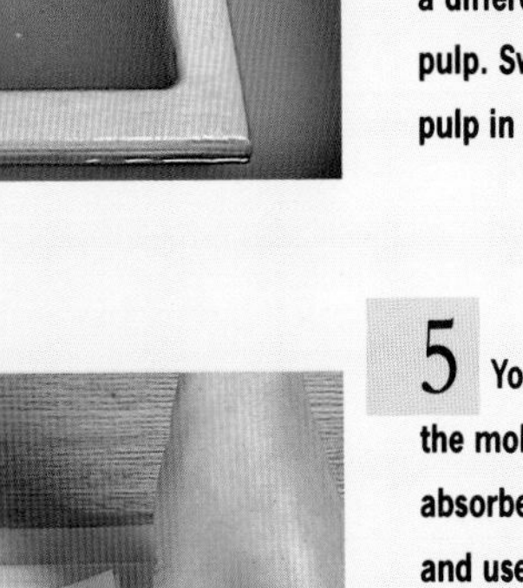

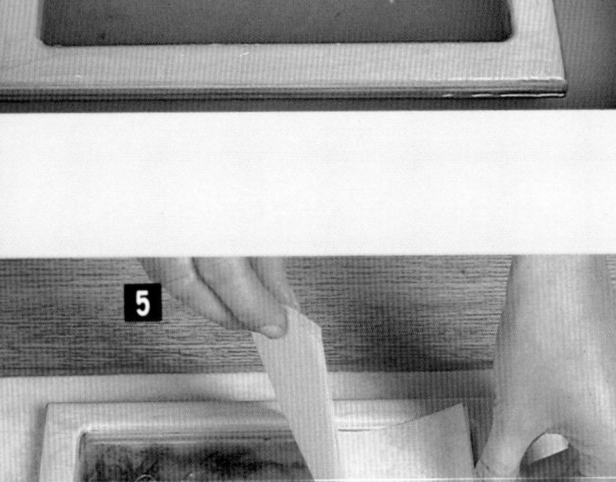

3 With the screen still submerged, pour in a small amount of a different-colored pulp. Swirl the second pulp in gently with the first and slap the coalescing pulp with the back of one hand to create an even overall layer.

4 Holding the mold level with both hands, lift it straight up and out of the water. Rest the mold over the vat, and let it drain for 5 minutes.

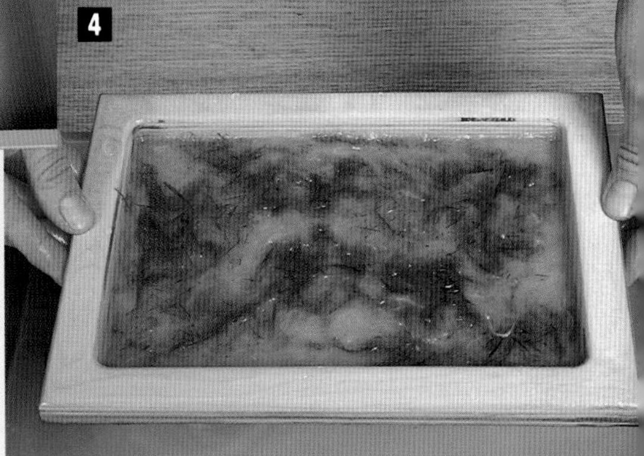

5 You can place the mold on an absorbent surface and use a couching felt to remove water.

6 Stand the mold upright and leave the paper to dry.

7 When the sheet is completely dry, gently rub the back of the screen to loosen the paper. You might need to use a small spatula to separate the sheet from the screen. Then carefully peel the paper from the mold.

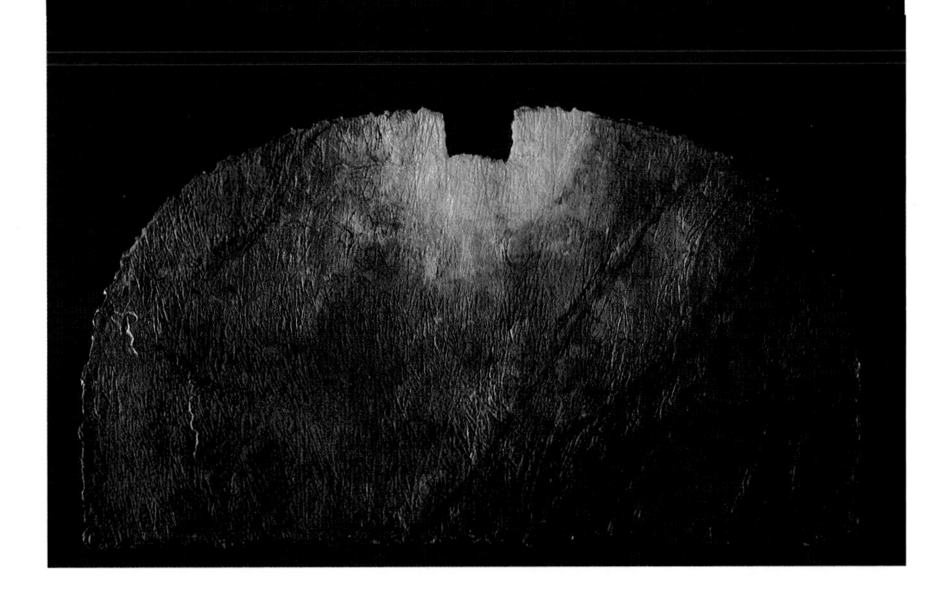

Soledad Vidal Massaguer: The Slow Way to the Light. Handmade paper of cotton and Spanish flax, dyed with natural pigments. 28×16in. (70×40cm).

New Outlines: making shaped paper

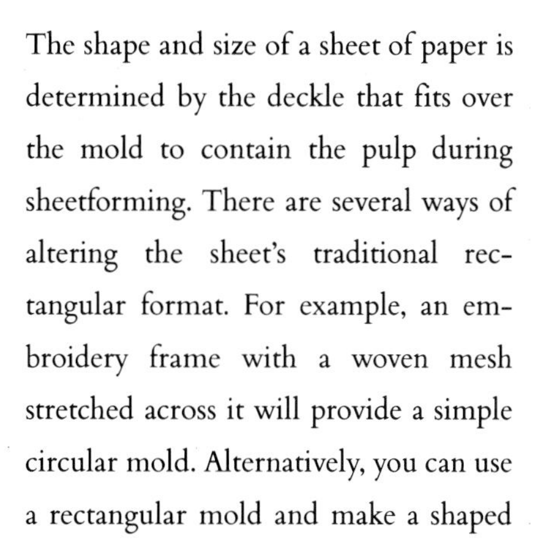

The shape and size of a sheet of paper is determined by the deckle that fits over the mold to contain the pulp during sheetforming. There are several ways of altering the sheet's traditional rectangular format. For example, an embroidery frame with a woven mesh stretched across it will provide a simple circular mold. Alternatively, you can use a rectangular mold and make a shaped deckle to fit over it. Simple shapes can be cut out of plywood using a jigsaw, and you can make more complicated deckles with a product called Buttercut (available from papermaking suppliers). An envelope-shaped deckle is useful when making stationery papers. Cookie cutters can serve as small, shaped deckles, and a heart-shaped cake mold makes a good Valentine deckle.

Suminagashi (Japanese marbling) on handmade paper (above).

Joan Hall: Foruja. Mixed medium on handmade paper. 7×9ft. (2×2.7m), (right).

PROJECT

Fun Shapes

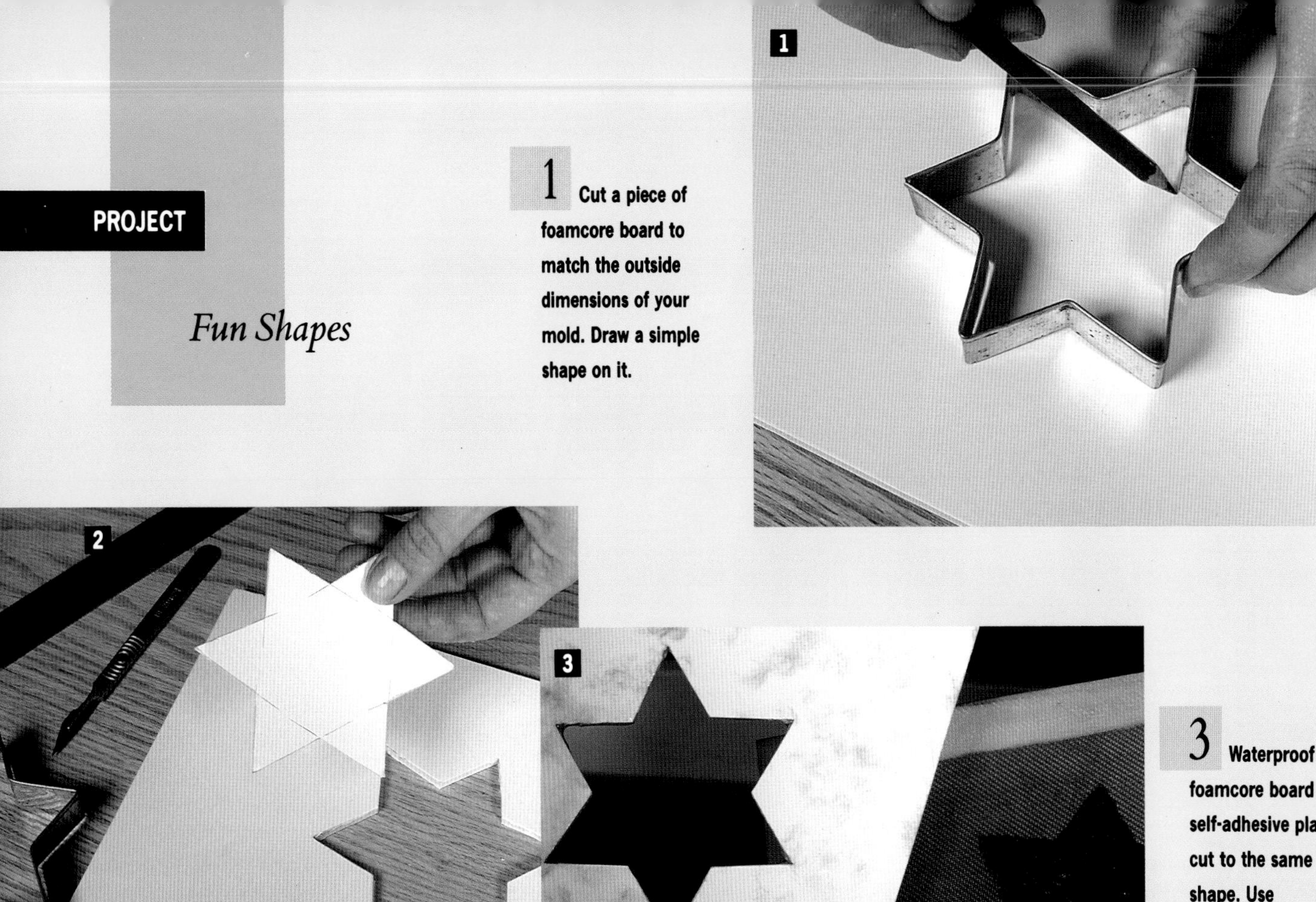

1 Cut a piece of foamcore board to match the outside dimensions of your mold. Draw a simple shape on it.

2 Carefully cut the shape out.

3 Waterproof the foamcore board with self-adhesive plastic cut to the same shape. Use waterproof tape around the edges before you immerse it in the vat. If you use wood for your deckle, coat it with water sealant.

4 Use the shaped deckle on the mold in the same way as a regular deckle. The paper will be the shape of the negative star in the deckle.

5 Laminate the star onto a base sheet.

YOU WILL NEED

- *prepared pulp*
- *mold*
- *couching cloths*
- *pressing boards*
- *foamcore board ("Buttercut")*
- *adhesive-backed plastic/waterproof tape*
- *water sealant*
- *pencil*
- *cutting board*
- *scalpel or craft knife*
- *cookie cutter shapes*

TECHNIQUES

- *sheetforming*
- *couching*
- *laminating*
- *pressing*

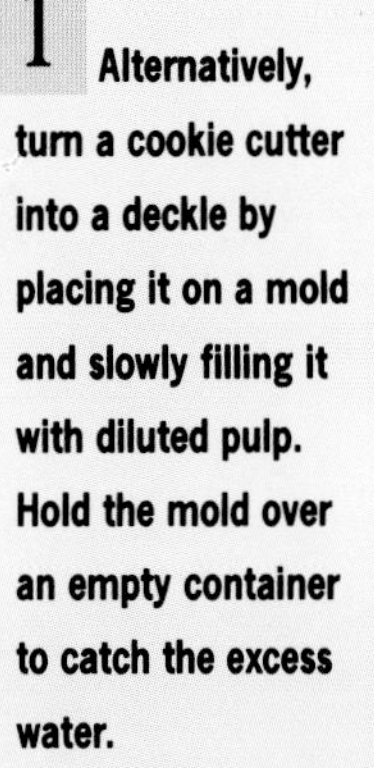

1 **Alternatively, turn a cookie cutter into a deckle by placing it on a mold and slowly filling it with diluted pulp. Hold the mold over an empty container to catch the excess water.**

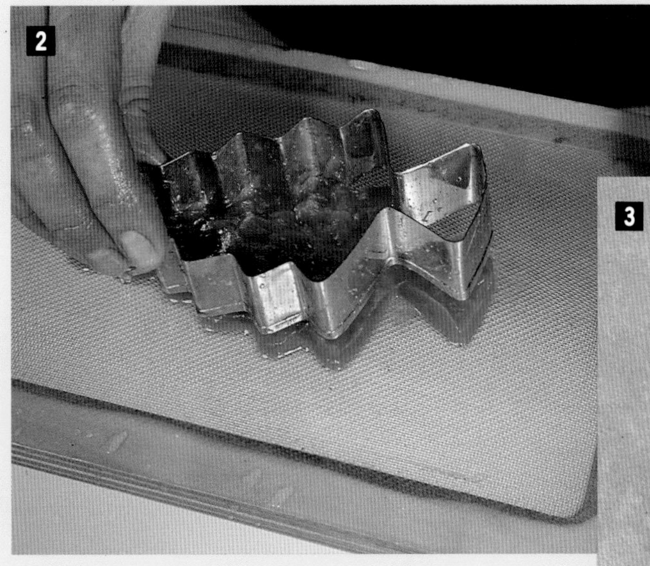

2 **When the pulp has drained, lift the cutter carefully off the mold.**

3 **You can couch the shape separately.**

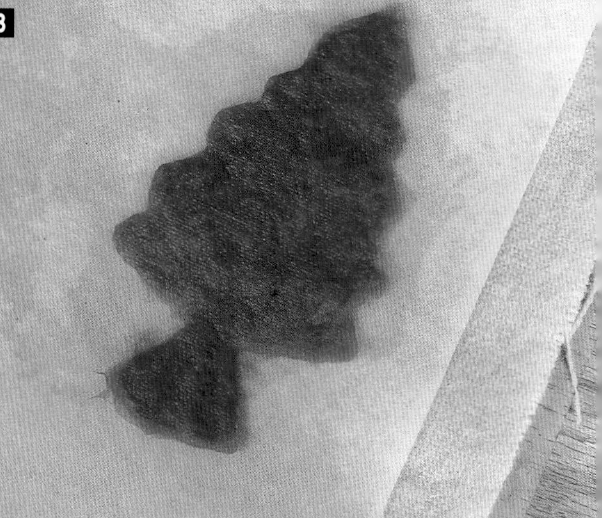

4 **A variety of shapes are possible in the colors of your choice.**

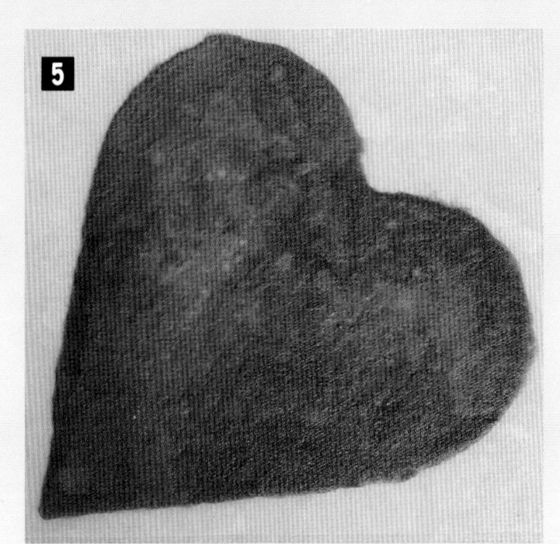

5 **You can also mix different color pulps to create special effects.**

Géza Mészáros: Cognitive and Affective Informations. Handmade paper, "Rost-relief": modeled and painted while the sheet is still wet. 86×145×13in. (220×370×35cm).

Embossing

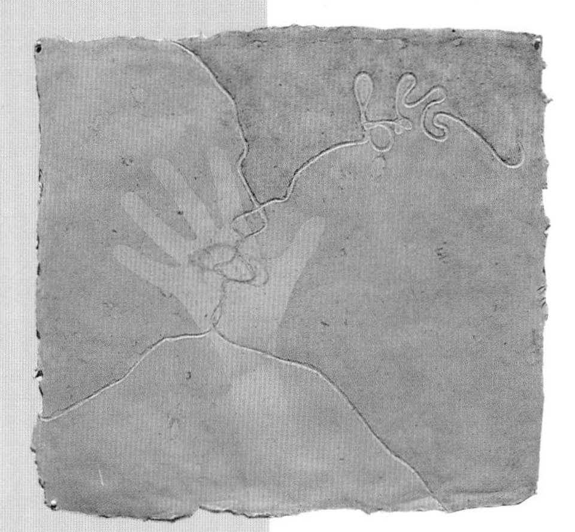

Embossing is a process which creates a raised or recessed image in a sheet of paper. The surface of the paper is altered by the pressure of a relief element applied against it. The heavier or thicker the paper, the greater the possible degree of embossment. Different levels of embossing can create a dramatic interplay of light and shadow.

You can emboss a sheet of paper either by forming it on or pressing it against a textured surface. Paper has a "memory" in this respect: it retains an impression of whatever surface it comes into contact with, from the mold it is formed on, to the felts it is pressed between, and the board it is dried against. Textured fabrics, such as lace, millinery net and embroidered cloths, leave a clear imprint when pressed against a wet, freshly couched sheet of paper. Wood blocks designed for textile printing provide a good surface for embossing. Or, make your own lino-block or collagraph. A sheet of paper that has dried on the mold will take a better impression than a sheet which has been couched and pressed before drying.

Julie Norris: The Last Oyster. Recycled gampi tissue with pastel. 17×18in. (43×46cm), (above).

Margaret Ahrens Sahlstrand: Fern and Salmonberry (detail). Embossed image on handmade paper. Plate size 30×40in. (76×102cm), (right).

TECHNIQUE

Embossed Shapes

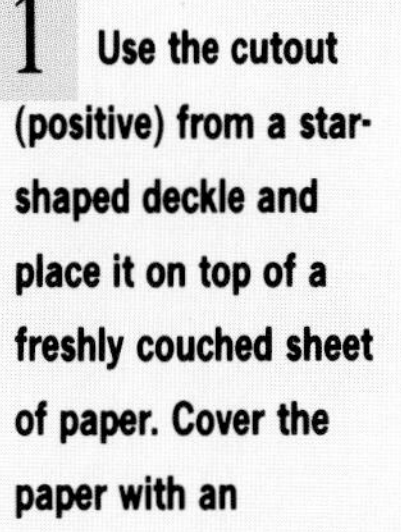

1 Use the cutout (positive) from a star-shaped deckle and place it on top of a freshly couched sheet of paper. Cover the paper with an interleaving felt. A thicker felt on top of this will provide a cushioning effect and help the paper to conform to the depth of the image.

YOU WILL NEED

- *prepared pulp*
- *mold and deckle*
- *couching cloths*
- *pressing boards*
- *scissors*
- *relief elements (string, textured fabric, lino-block, foamcore board)*

TECHNIQUES

- *sheetforming*
- *couching*
- *pressing*

2 When you are embossing, press each sheet of paper separately to avoid indenting other sheets. Remove the star carefully. The image has been embossed into the sheet.

1 Or, unwind a length of string and cut a piece off. Place it on a freshly couched sheet of paper. If it has been wound tightly, the string will often fall into a natural twist.

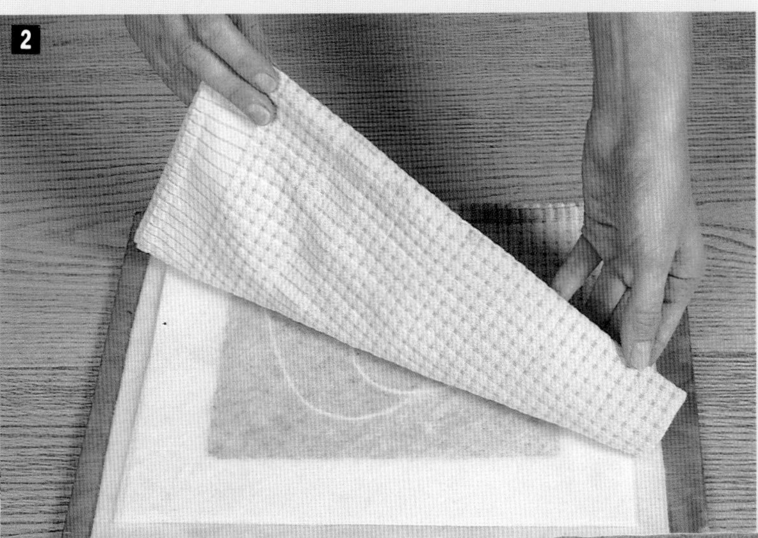

2 Cover the paper and string with felt, and then place a textured cloth on top of the felt. Press.

3 Both the string and the textured cloth have been embossed on the paper, creating different surface levels and an intriguing low-relief quality.

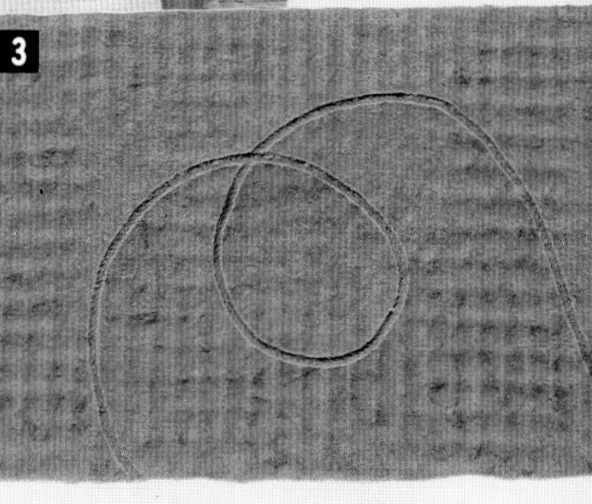

Gerry Copp: Tropical Fish Collage. Handmade recycled paper on wood. 19½×16in. (50×40cm).

Collage and Combining Techniques

Collage gets its name from the French verb *coller*, to paste. The art of collage consists of pasting together various materials, but particularly paper, to create a composite image. The histories of collage and papermaking are closely linked. The earliest known collages were made in the 12th century by Japanese calligraphers, who copied poems onto various pieces of tinted paper pasted together into sheets. Cutout paper shapes of birds, stars, and flowers were used to ornament the text.

Many contemporary collages are made from gathered scraps of used paper. They juxtapose disparate images and textures, torn and cut edges, hand dyed and painted papers. Newspapers, magazine pictures, advertising posters, wrapping papers, computer cards, colored and printed papers of all kinds are retrieved and given a new esthetic ▶

John Gerard: Bogen (Arch). Collage, handmade paper, in 13 parts. 71×46in. (180×116cm), (above).

Richard Flavin: Collage G-10. Collaged handmade paper with wood-block monoprint. 12×23½in. (30×60cm), (right).

PROJECT

Collage Shapes

1 **Cut a piece of foamcore board to match the outside dimensions of your mold.**

YOU WILL NEED

- *prepared pulp: 2 or more colors*
- *mold and shaped deckles*
- *couching cloths*
- *pressing boards*
- *wide, soft brush*

TECHNIQUES

- *sheetforming*
- *couching*
- *laminating*
- *shaped paper*
- *pressing*

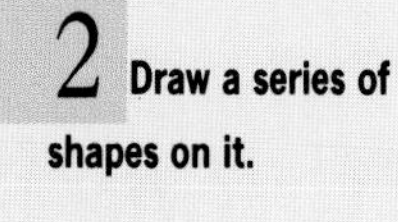

2 **Draw a series of shapes on it.**

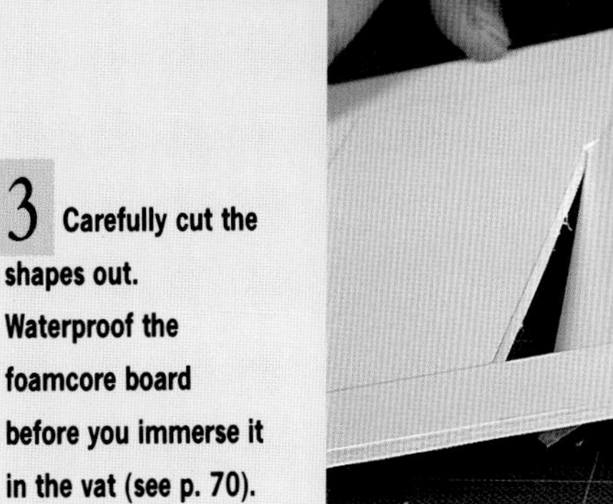

3 **Carefully cut the shapes out. Waterproof the foamcore board before you immerse it in the vat (see p. 70).**

4 **Make a base sheet, couch it, and set it aside while you are using the shaped deckles. Place the deckles on the mold, and make each shape with a different colored pulp.**

5 **Couch the shapes separately, each interleaved with felt in the order in which they will be laid. Put between boards under a light weight or in a press. Use just enough pressure to remove the excess water.**

◀ meaning. The tradition of collaging waste papers has recently been extended to include handmade paper, thus bringing together the permanence of craftsmanship with the disposability of mass production.

John Gerard: Collage. Handmade paper in 16 parts. 35½×35½in. (90×90cm), (left).

Richard Flavin: Collage B-4. Collaged handmade paper. 17½×17½in. (45×45cm), (right).

A COLLAGE COLLECTION

Start by building up a collection of papers, paying attention to their surface texture and tonal qualities as well as any printed text or imagery. Printed tissue fruit wrappers, weathered poster papers, metallic foil paper, coated papers, tickets, and torn postage stamps, for example, can form a useful basis for recycling and collage. Although recycled paper generally lacks the subtlety of a handmade sheet, it can possess a beauty and charm of its own in terms of color, feel, and texture.

Gerry Copp: Offcuts. Handmade recycled paper on wood. 27½×12in. (70×30cm), (right).

6

6 Lift the first shape from its felt.

7 Place it in position over the base sheet. Use a damp sponge to lift the shape if it appears fragile.

7

8

8 A wide soft brush can be used to smooth the shape down and tap out any air bubbles that may be present.

COMBINED COMPOSITIONS

There are various ways to incorporate throwaway scraps of paper or other ephemeral collage material into your handmade production process; or you might prefer to work exclusively with your own handmade papers. You can combine the techniques of laminating, embedding, embossing, and creating shaped papers, shown earlier in this section, to make a collage composition. For example, textured pieces of paper can be laminated onto freshly couched sheets of a contrasting color and combined with embossing. Or an embedded image might be revealed by couching a decorative top sheet that has a sprayed pattern of open spaces.

9 **Continue to laminate each shape onto the base sheet until all the pieces are in place.**

10 **Return the finished sheet to the press and use a heavier weight or increased pressure to secure the bond**

9

Otavio Roth: Untitled. 50,000 paper elements made from chewed cotton rag.

Paper's Precursors

Tapa cloth is mostly undecorated, but can also be stamped, painted, or stenciled with geometric patterns and animal and plant motifs.

A number of ancient materials resemble paper, but are not, strictly speaking, paper at all. Some served as a writing surface, and some had additional functional and spiritual uses. Papyrus, *tapa* and *amate* (from tree bark), and rice paper can all be called precursors of paper with regard to their uses, and all are derived from vegetal material. Their manufacture, however, differs from that of true paper: they are not made from macerated fibers that have been dispersed in water and collected on a mold.

PAPYRUS: THE ANCESTOR OF PAPER

Papyrus is made from the stems of the papyrus plant (*Cyperus papyrus*), which used to grow prolifically in the marshy areas along the banks of the Nile. The oldest surviving specimen of papyrus dates back almost 5,000 years to 3,000 B.C. The papyrus plant was used for an astonishing number of purposes, including fuel, boats, ropes, mats, sandals, blankets, and clothes. Its chief function, however, became that of a writing material. Papyrus sheets were cut to standard sizes and polished with pumice. They were used either singly or glued together to form a longer roll. (The Great Harris Papyrus manuscript, preserved in the British Museum, is 133 feet/40.5 meters long.)

POLYNESIAN PAPER

Tapa is the Polynesian word for a cloth-like material made from the inner bark of a variety of trees, especially those in the mulberry family. It is produced by soaking the bark and pounding it into sheets. Various forms of tapa were used throughout the Pacific, South and Central America, Africa, and Southeast Asia for many purposes. In ancient Hawaii, ▶

PROJECT

Making a Papyrus sheet

Successful papyrus-like sheets have been made using substitute fibers from the bulrush stem, cornstalk, and even the amaryllis plant. Find out whether there are any such plants growing in your area that you could harvest to make your own papyrus paper. Alternatively, dried strips of papyrus and papyrus-making kits are available from the Papyrus Institute in Cairo or from hand papermaking suppliers. The kit contains two sets of dried pre-cut strips, 5in. (12.5cm) and 6in. (15cm) long.

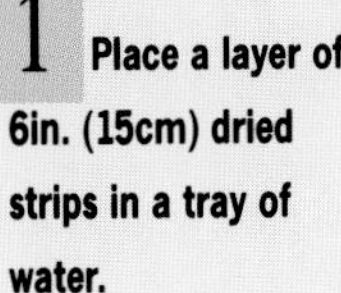

1 **Place a layer of 6in. (15cm) dried strips in a tray of water.**

2 **The 5in. (12.5cm) strips should be laid on top. Soak at room temperature for 2-3 days.**

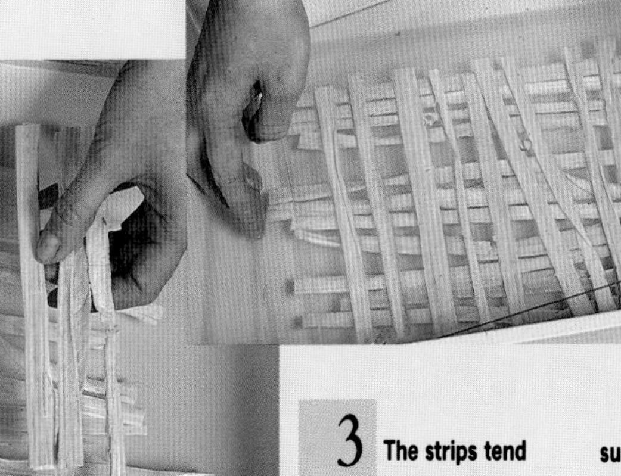

3 **The strips tend to float, so it is best to devise a weight, such as a sheet of glass, to keep them submerged.**

4 **Take the strips out of the water and roll them on a smooth board with a rolling pin. Re-immerse the strips in a tray of fresh water for a further two days. Having absorbed some water, the strips are less likely to float, and you may be able to dispense with the weight.**

YOU WILL NEED

- *collected plants (as recommended)*
- *sharp knife (for harvesting stems and slicing into strips)*
- *vegetable peeler (for removing skin from stems)*
- *or papyrus kit*
- *shallow tray*
- *wooden board*
- *rolling pin*
- *2 cotton towels*
- *2 cotton handkerchiefs*

tapa was used for ceremonial clothing and burial wrappings, bandages, bedding, string, and lampwicks.

AZTEC PAPER

Amate is also a pounded bark paper. It was originally made by the Aztecs from the inner bark of wild fig trees, although the Maya had earlier developed a beaten bark material called *huun.* The Aztec word *amatl* means both paper and fig tree. Amate is made by cooking the strips of inner bark with a wood ash lye. The strands are rinsed and laid out in a grid formation on a wooden board. The fibers spread as they are beaten and fill the gaps. The amate sheets are left to dry on the boards in the sun.

Like *huun* before it, *amatl* was widely

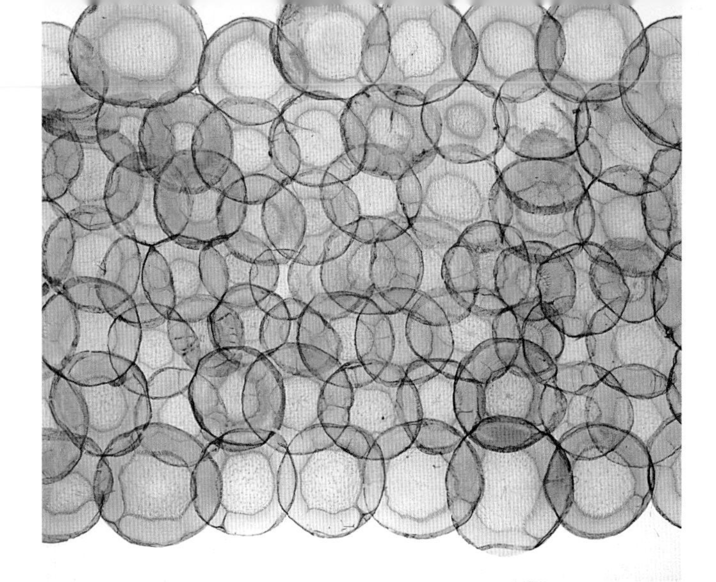

Translucent carrot paper made from sliced vegetable cross-sections laminated in a process similar to the manufacture of papyrus.

5 After the second two days, repeat the rolling process, using a little less pressure. Return the strips to the water (without changing it) for a further three days. When the strips become translucent and sink to the bottom of the tray, they are ready for sheetmaking.

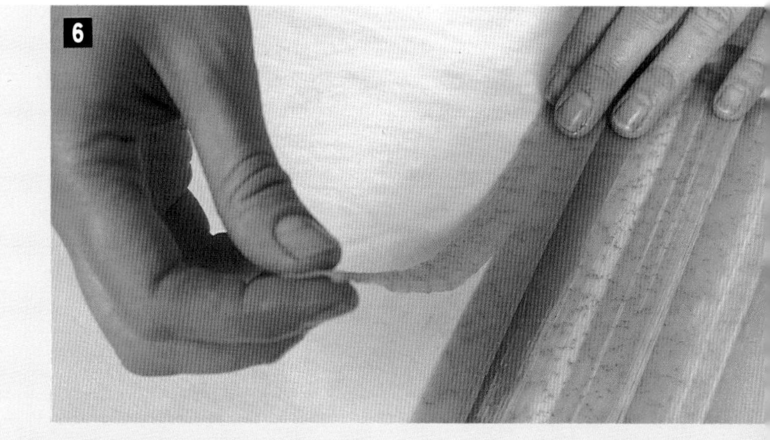

6 Lay a small towel on a flat surface and cover it with a piece of cotton, such as a clean handkerchief. Take one of the wet strips and lay it in a vertical position along an edge of the cloth. Place another strip of the same length parallel to the first with a slight overlap of about $\frac{1}{16}$in. (1-2mm). Continue to lay out the remaining strips of the same length in this way.

7 Lay the other strips in a horizontal direction on top of the first layer, overlapping each strip as before.

used as a writing surface for sacred calendars, prophecies and divinations, historical records, and medical treatises. It was also central to the religious practices of the Aztecs. Amatl is still made by the Otomi Indians of southern Mexico, who have preserved its powerful symbolic role in their ceremonies.

RICE PAPER

Rice paper is the erroneous name for a paper-like material made by cutting the spongy inner pith of the rice paper plant (*Tetrapanax papyriferus*) into fine veneer-like sheets.

UNUSUAL PAPERS

Experiments in the 18th century to find new materials for papermaking led to the production of assorted "strange" papers. Naturalists tried working with an extraordinary number of alternative plant and vegetable fibers, including seaweed, pine cones, moss, nettles, potatoes, grass, grapevines, and straw.

UNUSUAL PAPERMAKERS

The way that wasps make their nests from chewed-up wood pulp, and the paper-like product of their labors, were noted by the French naturalist and physicist René-Antoine de Réaumur in 1719. In modern times the wasps have found an imitator in Brazilian artist Otavio Roth. His techniques include actually chewing a piece of his cotton T-shirt for two hours to produce a mouthful of pulp.

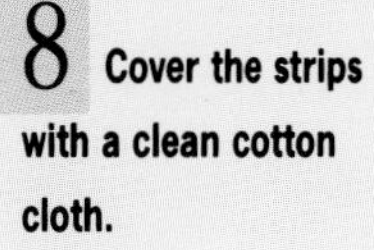

8 **Cover the strips with a clean cotton cloth.**

9 **Cover the cotton cloth with a towel.**

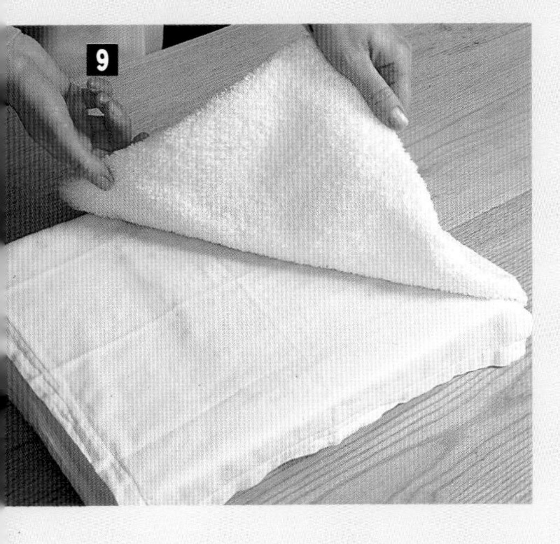

10 **Carefully roll the "sandwich" of papyrus strips to press out the excess water. Remove the wet towels, but leave the pieces of cotton cloth on each side of the papyrus sheet, and place it between a thick layer of folded newspaper. Put a board and weight on top of the newspaper or place it between wooden boards in a small press, if you have one. Change the wet newspaper after 2 hours, then after 6 hours, and again after 24 hours.**

Leave the papyrus sheet in the press until it is completely dry, continuing to change the newspapers every 24 hours. Finally, remove the papyrus from the pieces of cotton cloth and weight the sheet between clean, dry paper for a further 48 hours.

Watermarks are formed in the paper as it is made, which makes them difficult to counterfeit. As a result, they are an important feature of banknotes and other security papers.

Watermarks: a hidden image

A subtly shaded chiaroscuro watermark portrait of Nicholas Louis Robert, inventor of the first paper-making machine.

A watermark is a translucent area in a sheet of paper, usually in the form of an outlined image or initial, which is barely visible except when held to the light. It is the result of an impression left in the pulp during formation of the sheet by a fine wire shaped into the desired design and attached to the surface of the mold. The pulp settles in a thinner layer over the slightly raised area of the watermark device. The earliest examples of watermarks come from late 13th-century Italy, and the oldest known watermark is a rough cross with a small circle at the end of each branch and a larger one in the middle.

The term watermark is a misleading one, since it is the wire design that creates the translucent areas in the sheet and water plays no particular part in the production of the image. The French word *filigrane* and earlier English terms "wire-mark" and "papermark" are more accurate descriptions of the hidden images which are a distinctive feature of European paper.

MAGICAL IMAGES

A tremendous variety of pictorial designs, including images of animals and birds, humans, flowers and trees, heraldic themes, and simple shapes such as circles, crosses and stars, can be found among early watermarked papers. The exact purpose of these images is unknown, but they probably helped identify the molds of illiterate papermakers. Some theories suggest that they may have had a religious or magical significance. Today, watermarks serve mainly as the logo or trademark of a particular paper, mill, papermaker, or artist.

PROJECT

Creating a Watermark

1 Draw a simple outline on a piece of paper. Then bend a wire thread with a small pair of pliers to the shape of your design. Components of a more elaborate wire design need to be soldered. If you use fishing line, sew it directly onto the mold cover.

YOU WILL NEED

- *pencil*
- *paper*
- *soft brass wire (.030–.040 gauge) or other rustproof wire, available from jewelry suppliers*
- *fishing line (alternative to wire)*
- *small pliers*
- *needle and thread*
- *hand soldering iron (optional)*
- *mold and deckle*
- *prepared (short fiber) pulp*
- *couching cloths*
- *pressing boards*

TECHNIQUES

- *sheetforming*
- *couching*
- *pressing*

2 Sew the wire shape onto the surface of the mold with a strong, fine thread. You need only attach it at key points to keep it firmly in place. Remember that the design will appear as a mirror image on the sheet of paper.

3 Use a short to medium length fiber for your pulp to achieve a clear image. As you raise the mold from the vat, less pulp will settle over the slightly raised mark than over the rest of the screen, and the watermark will form as part of the sheet. When the sheet is pressed and dry, hold it up to the light. The more translucent areas of the sheet, where the watermark image was attached to the mold, will become visible.

PROJECT

Experimental Watermarks

There are other ways to create a watermark effect. Small adhesive labels can be applied to the surface of the mold when it is dry. A water-resistant adhesive, or even nail varnish, can be used to "draw" a design on the mold - but you will not be able to remove it from the screen afterward. Water cannot drain through the masked areas during sheetforming and so the pulp does not settle on them.

YOU WILL NEED

- *adhesive labels or water-resistant adhesive*
- *prepared (short fiber) pulp*
- *mold and deckle*
- *couching cloths*
- *pressing boards*

TECHNIQUES

- *sheetforming*
- *couching*
- *laminating*
- *pressing*

1 Attach adhesive labels to the surface of the mold.

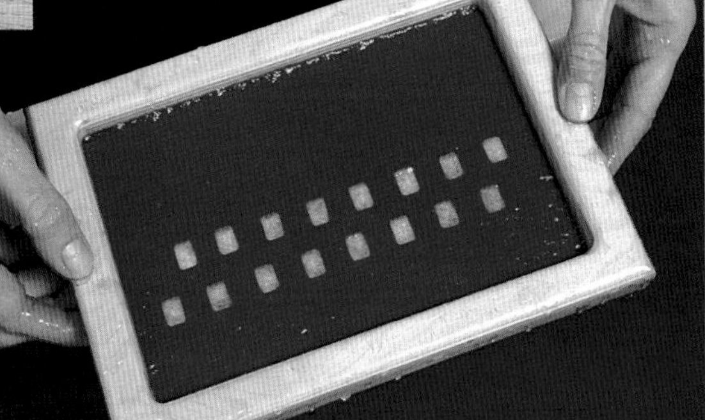

2 Pulp will only settle where the water is able to drain through the screen.

3 Laminating a sheet with this watermark design onto a freshly couched sheet of a different color creates a see-through effect.

4 The base sheet can be seen through the design areas that were masked by the labels.

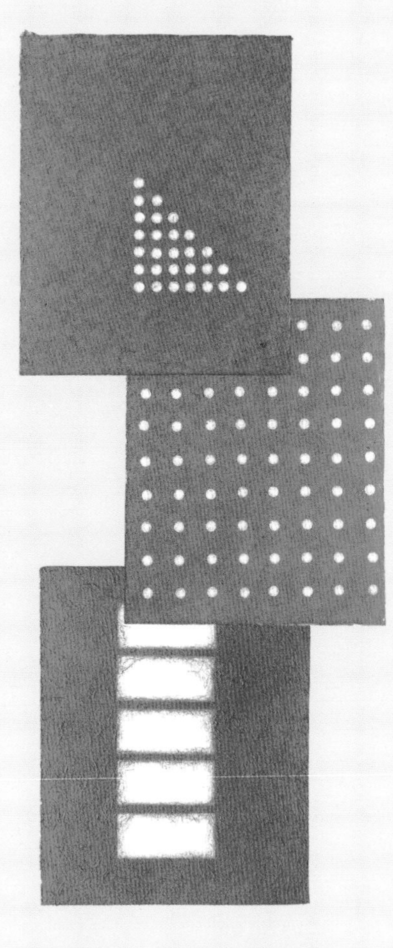

These designs were formed by attaching adhesive labels to the surface of the mold before sheetforming.

Golda Lewis: Citigeometric. Paper made directly on unsized canvas with assemblaged materials. 30×36in. (76×91cm).

Pulp Painting

Géza Mészáros: The Big Iconoclastic Icon. Handmade paper, with gold paint. 61×46in. (155×115cm) (above).

Pulp painting uses paper pulp as if it were paint. The pulp is usually applied to a supporting surface, which may itself be paper, using various freehand techniques. It can be sprayed, spattered, squeezed, daubed, dripped, and poured.

Many artists have been fascinated by the inherent qualities of paper: its malleability, its color and texture. They have learned to make paper and to make paper work for them as a tool in developing their art. Their work often reveals more about the nature of paper itself than about paper as a supporting surface for other media. Approaches to working with pulp result to some extent from the particular interests of the artist and frequently combine drawing, painting, and printmaking with one or more aspects of the papermaking process.

The sequences involved in traditional methods of Eastern and Western sheetforming (see p. 38-9) and watermarking (see p. 82) offer a strategy for creating ▶

Kathy Crump: Fan Shape (detail). Pulp painting, folded, hibiscus over kozo paper. 18×25in. (46×63cm), (right).

TECHNIQUE

Pulp Painting

1 **Make a sheet of paper and rest the mold over a shallow tray to drain. Place a wire frame, such as a cake cooling rack, on top of the sheet. Prepare the colored pulps with formation aid and put them in plastic squeeze bottles. Test to see that the pulp flows evenly through the nozzle.**

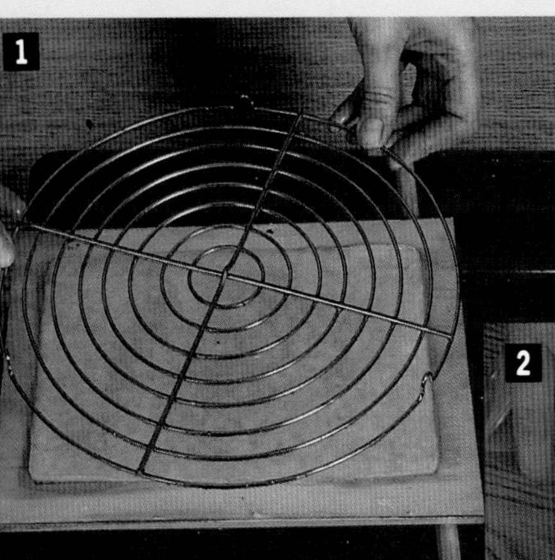
1

2 **Apply the first colored pulp, directing the flow of pulp into a section of the cooling rack. Squeeze the bottle slowly with a steady pressure and paint in a thin layer of pulp. Wait for the water to drain before applying any more pulp, or it will spill over into the adjacent section.**

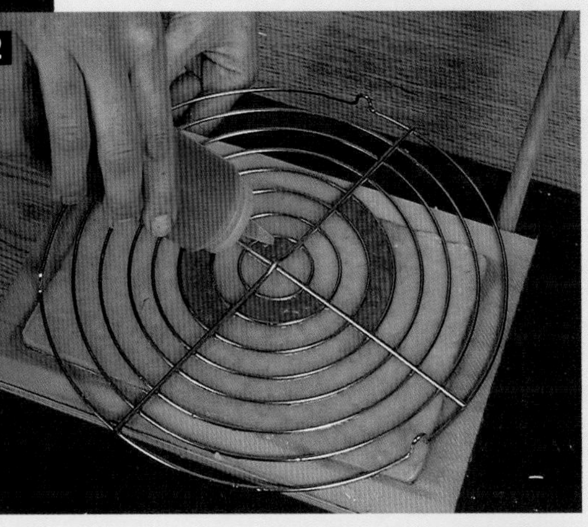
2

3 **Gradually fill in the sections using one colored pulp at a time.**

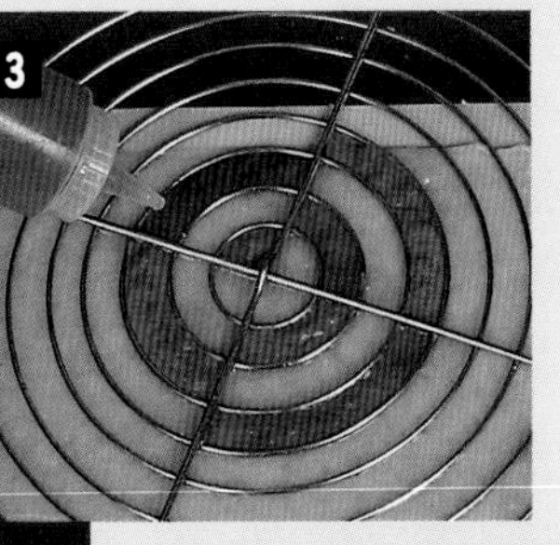
3

4

4 **When the water has drained, carefully remove the frame, and then couch and press the sheet.**

pictures. The Japanese method of multiple dippings, for example, can be used to reimmerse the mold into vats of different-colored pulps.

PULP PAINTING TOOLS

Plastic squeeze bottles can be used to hold the pulp for painting. Almost any well-diluted pulp can be used, although better results are achieved with a longer, over-beaten fiber (see p. 102) combined with a formation aid. Abaca and flax are recommended, especially where translucency or intense color is desired. The addition of formation aid helps prevent the fibers from clumping together and blocking the nozzle of the bottle. The proportions of water, fiber, and formation aid can be adjusted until you have a pulp that flows easily. It is better to start with a little formation aid and add more as necessary.

A template or stencil can be used as a guide when pouring the colored pulp. Flexible plastic strips can be positioned as barriers, and brass or copper "fences" can be shaped to contain the poured pulp. When the barriers are removed, a slight shake before the water drains will soften and merge adjoining borders of pulp. Alternatively, you can apply the pulp freehand using spontaneous painterly strokes.

TECHNIQUE

Painting Freehand

You can also couch a base sheet onto a dampened felt and paint pulps into it. If you have a working drawing, the base sheet can be marked lightly with a plastic knife in accordance with your design. If you are layering colored pulps, make a note of the order in which to apply them. Place a thicker felt underneath everything to absorb the water from the pulps.

1 **Apply the first colored pulp in a thin, steady stream, working spontaneously or following your design.**

2 **When the water has drained, proceed with the second color.**

3 **Continue applying each color in turn, making sure that the water always drains away between applications.**

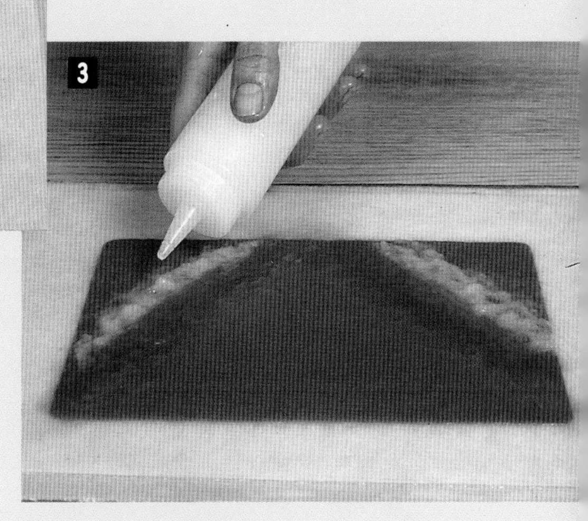

YOU WILL NEED

- *wire frame (cake cooling rack)*
- *plastic squeeze bottle or turkey baster*
- *prepared pulp*
- *formation aid*
- *mold*
- *catch tray or couching felt on board*

TECHNIQUES

- *sheetforming*
- *couching*
- *pressing*

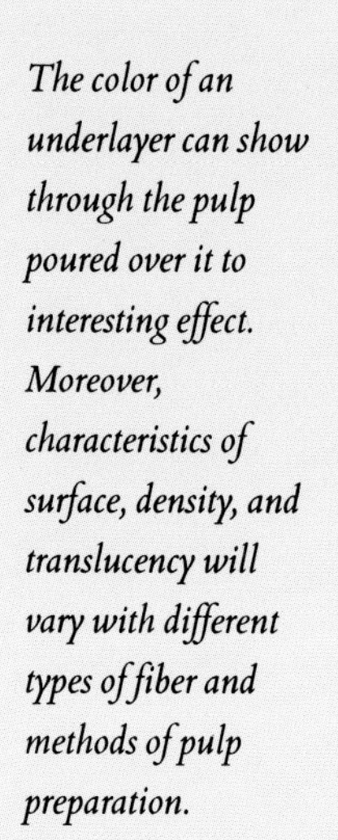

The color of an underlayer can show through the pulp poured over it to interesting effect. Moreover, characteristics of surface, density, and translucency will vary with different types of fiber and methods of pulp preparation.

Elaine Koretsky demonstrating pulp spraying techniques at an International Paper Conference.

Pulp Spraying

Donna Koretsky: Tree Rings: expanded version (detail). Sprayed flax fiber. 78½×23½in. (200×60cm).

Pulp spraying is a popular technique which can be used to create large sheets of paper without a large vat system. The device for pulp spraying is based on a standard industrial tool for spraying paint, plaster, or concrete, and is powered by an air compressor. The effect is similar to that of an airbrush, and subtle color changes can be produced by gradually building up and overlapping successive layers of pulp. Pulp spraying can be successfully used on both two- and three-dimensional surfaces.

Pulp spraying equipment consists of a spray pistol, a hopper (container), and an air compressor of at least one, and preferably two, horsepower. The size of the compressor determines the volume you can spray. The beaten pulp is poured into the hopper, which is linked to the spray pistol by a length of hose. Another hose connects the pistol to the air compressor. When the air compressor has built up pressure in the tank, the trigger on the pistol is squeezed to release the pressure and spray the pulp. The pistol comes with interchangeable nozzles which offer a variety of wide and narrow spray patterns. You can obtain a complete system with full directions for use from a hand papermaking supplier.

CHOICE OF PULP AND SURFACE

You can spray almost any pulp, with the addition of formation aid (see p. 27), onto a variety of porous surfaces. Abaca is a good pulp to start with if you want to form strong, thin, translucent sheets. Cotton pulp can be used to build up an opaque surface. Longer fibers can be prepared to produce a pulp which will give a strong, high-shrinkage paper for sculptural work. You can use three-dimensional molds or armatures to ▶

Two-dimensional Spraying

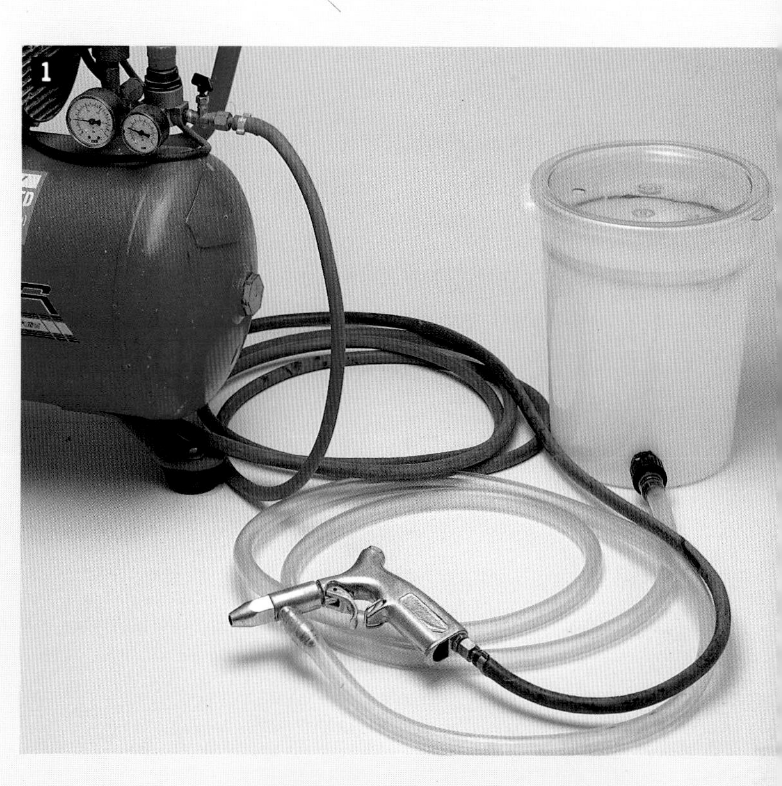

1

1 Connect the spray gun to the air compressor. Drain the prepared pulp for a few seconds and add $\frac{1}{2}$-1 cupful of formation aid to each gallon (4.5 liters) of drained pulp, stirring it in well. The pulp mixture should feel quite slimy. Have a bucket of fresh water standing nearby to dilute the pulp as required. Sizing and other additives (see p. 27) may also be mixed in and sprayed with the pulp. A dilute mixture of methylcellulose in the pulp will give extra rigidity to the finished piece. A small amount of acrylic gel will also act as an adhesive and will lend a reflective quality to the surface of the work. Wet the support fabric by spraying it with water. This will help the layers of diluted pulp to stick to it more securely.

YOU WILL NEED

- *pulp sprayer (available from leading hand papermaking suppliers)*
- *air compressor ($1\frac{1}{2}$-2hp)*

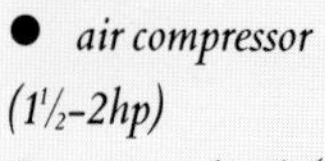

- *prepared pulp(s)*
- *formation aid*
- *woven fabric: unprimed cotton or linen canvas, silk screen fabric, muslin, etc.*

2

2 Then fit the pistol with a nozzle liner and connect it to the hopper with the hose provided.

3 Fill the hopper with the pulp and place it on a level surface near the piece you are going to spray. Start the air compressor, and when the correct working pressure has built up, pull back the trigger on the pistol to spray the pulp.

3

◀ create sculptural forms, and most woven material, provided it is not too loosely structured, is suitable for two-dimensional work. When dry, the mold or supporting surface can, if you wish, be removed from the sprayed paper form.

You will need to vary the pulp mixture to suit your surface and the nature of your project. A smooth surface, such as a silk-screen, will require a more dilute mixture than a coarser woven cloth. Because of the amount of water involved, spraying the pulp horizontally will minimize the risk of its sliding off a smooth surface. A wet-dry vacuum cleaner with a squeegee (rubber blade or roller) attachment can be used to speed the removal of excess water from the back of such a surface and will help to draw the pulp more securely onto it. Other flat or relief pieces can be sprayed from any direction. Spray the pulp in a succession of thin layers, letting it drain thoroughly between each application, so that it does not slide off. About 4–6 layers will produce a strong enough sheet, which will peel away from the support fabric when completely dry.

Stretch your chosen material taut, preferably across a frame as if you were making a standard mold cover (see p. 22). This applies to both flat and three-dimensional structures, because pulp will run and pool into any sagging areas. Suitable materials include unprimed cotton or linen duck, muslin, interfacing, silk-screen fabric, a close-weave screen wire, or even plastic sheets. Spraying pulp onto a nonporous material, such as a plastic drop cloth, will give a smooth and glossy surface to the sheet which dries against the plastic.

4 Spray the pulp in thin, even layers, waiting for the water to drain after each layer. Use the smallest nozzle liner to spray a narrow pattern. You can reduce the air pressure to spray as close as 2in. (5cm) from the work, but be careful not to spray too hard, or you may blow a hole in the piece. If the pistol becomes clogged, place a finger on the nozzle and gently pull the trigger until the pulp blows back down the hose. You may then need to adjust the ratio of formation aid and water in the pulp.

5 You can apply different colored pulps in solid bands or in diffused overlapping layers. Careful color blending can be used to achieve a subtle intensity of color which appears to radiate from within the piece. There is also considerable potential for combining techniques with the pulp sprayer, adding details by laminating, embedding, or applying pulp freehand with the aid of a plastic squeeze bottle. You can use a variety of stencils to mask out areas of the work while spraying.

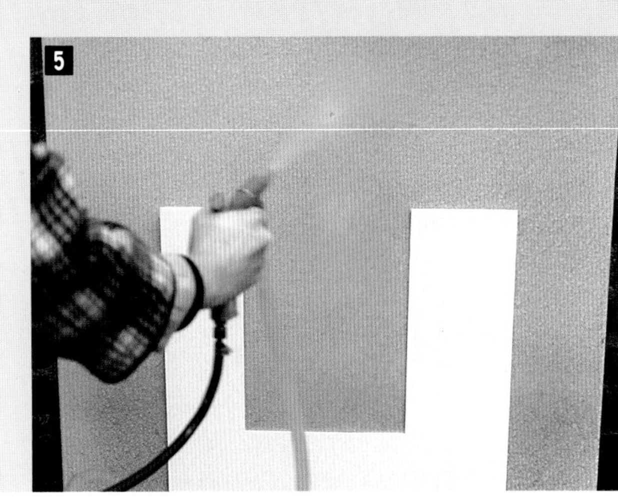

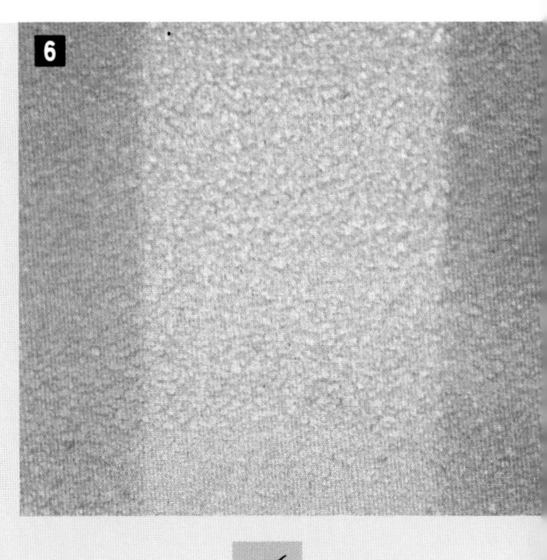

6 One of the characteristics of pulp spraying is this slightly textured pattern of water drops. After you have finished using the pulp sprayer, rinse the system through with clear water. Take out the nozzle liner, and remove any bits of trapped pulp.

A vacuum table is a useful piece of studio equipment and is relatively simple to construct using a wet-dry vacuum cleaner.

Vacuum Forming

It can be used when fragile or dimensional embedded elements prohibit traditional pressing; for casting; or for de-watering large sheets that cannot be accommodated in a press.

Vacuum-forming techniques offer an exciting alternative to traditional sheetforming methods and facilitate a wide range of paper pulp applications. A number of vacuum systems have been designed since the first vacuum tables appeared in the late 1970s, but all of them work on the same basic principle: atmospheric pressure is used to compress a newly formed layer of pulp, thereby extracting the water from it.

THE VACUUM TABLE

The system consists of a table, through which water is drawn from the pulp, a holding tank for the water, and a vacuum pump. The traditional table is made of marine plywood with small holes drilled at 6-10in. (15-25cm) intervals. Beneath the tabletop the holes lead into a vacuum chamber which is connected to the holding tank by a plastic hose. Another hose links the holding tank to the vacuum pump. The construction must be airtight to operate efficiently, and all surfaces, inside the vacuum cavity and on the table, are sealed with several coats of polyurethane varnish. An optional raised edge can be attached to the four sides of the table to contain the wet pulp if large amounts are used.

To create the vacuum, a plastic sheet is laid over the wet pulp, and a seal is formed between the plastic and the table surface. When the vacuum pump starts, the air is drawn out from under the plastic and within the vacuum chamber. As the air is removed, the wet pulp is compacted by the pressure of the atmosphere on the outside of the plastic, and water is extracted from the pulp.

Vacuum systems are frequently used for sheet sizes that exceed the limits of ▶

PROJECT

Vacuum forming

A packaged wet-dry vacuum system comes with an open-weave backing material, or space mesh, in two pieces, and a finer mesh polyester screen. Together they create the surface for the vacuum work and a space for water removal. A 2½in. (63mm) nozzle attachment is provided for the wet-dry vacuum hose, together with a roll of waterproof sealing tape. Two pieces of Ultra felt and a plastic drop sheet are also included.

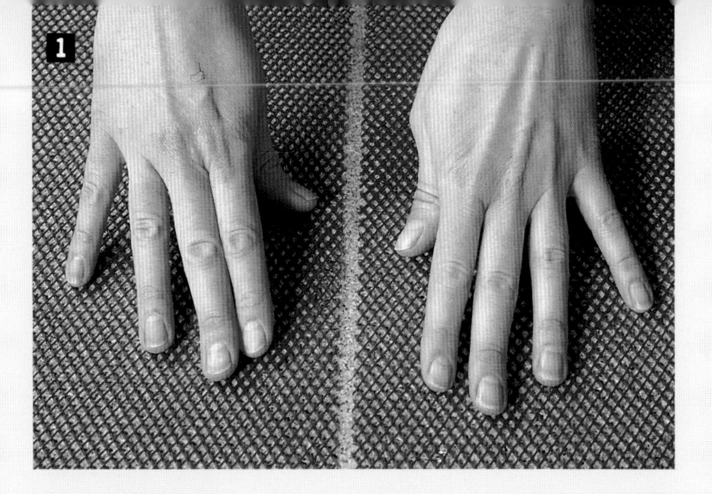

1 **Fit the two pieces of backing material together on a waterproof work surface, which should be 3-4in. (7-10cm) wider all around than the screen.**

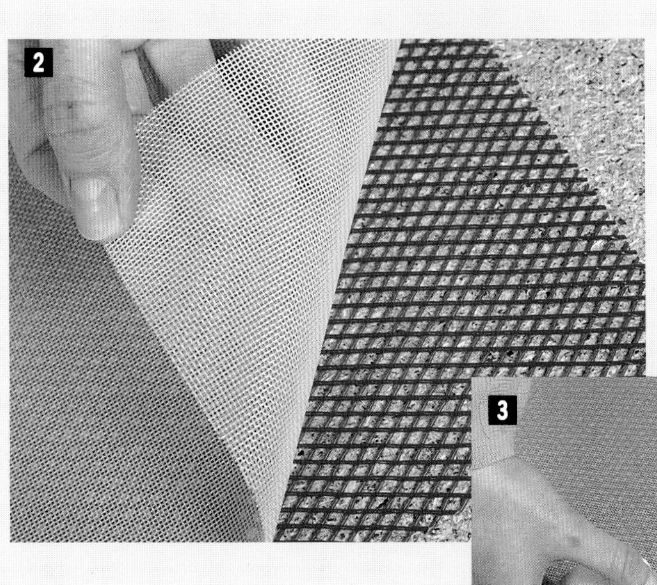

2 **Place the finer mesh screen on top of the backing material.**

3 **Position the nozzle at one corner of the screen, and seal it to the table and screen with the tape. Your working surface and screen must be completely dry to guarantee maximum adhesion of the tape.**

YOU WILL NEED

- *waterproof table surface*
- *aqua-vac — wet-dry vacuum system — (can be rented)*
- *hose attachment (available from leading hand papermaking suppliers)*
- *waterproof tape*
- *2 layers screen/mesh: spacer mesh/polyester screen (available from leading hand papermaking suppliers)*
- *couching felt*
- *plastic drop cloth*
- *prepared pulp(s)*

TECHNIQUES

- *sheetforming*
- *couching*
- *paper casting (optional)*
- *pulp painting (optional)*

4 Cover the screen with a layer of Ultra felt to support the work if you wish to remove it from the table after the vacuum operation. Dampen the Ultra felt before you lay it down or begin work on it. The vacuum table is now ready.

5 To see how the vacuum process works, first position a freshly made piece of paper on the Ultra felt.

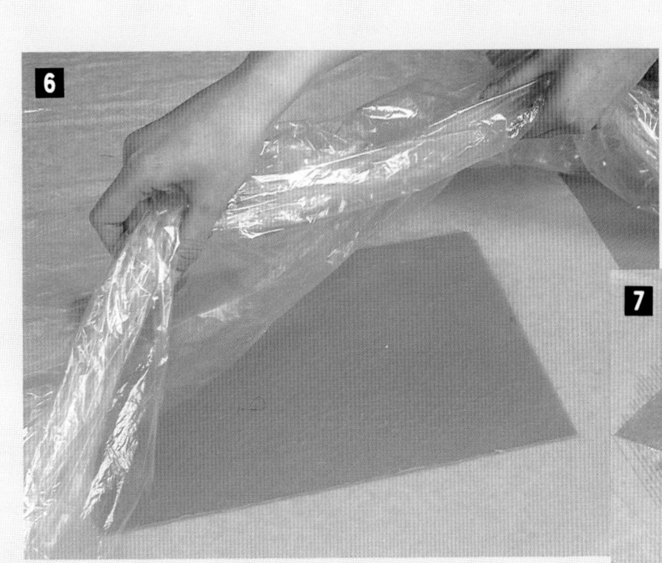

6 Cover the paper with the plastic sheet, making sure that the exposed edges of the table are wet. This is essential to achieve a good seal with the plastic. Any leaks or holes in the plastic will prevent the formation of a true vacuum.

7 Attach the wet-dry vacuum hose to the nozzle and turn on the cleaner. When the air beneath the plastic has been removed and atmospheric pressure bears down, you will see the water begin to be drawn out toward the nozzle attachments.

8 Continue the vacuum operation until all the water which can be extracted with the pressure available has been sucked out. Remove the plastic sheet before you turn the vacuum system off.

9 When you have finished, empty the wet-dry vacuum cleaner, let the screen mesh material dry before storing it (as flat as possible), and check that there are no holes in the plastic sheet.

the traditional paper press. The creation of large sheets, embedding and lamination, embossed forms, works in relief, collage, and pulp painting can all be effectively accomplished using a vacuum table.

BUILD YOUR OWN SYSTEM

You can make your own vacuum system by using a wet-dry vacuum cleaner. A quality wet-vac develops 7-10in. (17-25cm) Hg vacuum, which is about one third of the vacuum pressure of a deep vacuum pump. Probably the best approach to begin with is to borrow or rent the most powerful wet-dry vacuum cleaner available and order the smaller component parts from a papermaking supplier. A basic system provides enough materials for a work surface of 40×36in. (100×90cm).

The system can be used on any flat waterproof surface, such as a wooden or Formica table top, or even a floor covered with a thick sheet of plastic. You may wish to make a frame, using strips of wood, to fit around the edge of the surface area. This will stop the water from running off the table and possibly flooding your work space. Do not forget to waterproof the frame and seal the seams where it is joined to the table.

VACUUM TABLE TECHNIQUES

The vacuum table can be used for a wide variety of techniques. You can make a sheet of paper in a vat and couch it onto the Ultra felt as a base sheet for pulp painting techniques. Or you can work directly on the Ultra felt, perhaps sketching your design on it first. You can drain some of the excess water from your pulp and spread it onto the vacuum table to create a rougher sheet. You can manipulate the pulp into different forms or embed different materials into it, concealing them with further applications of pulp or leaving them partially exposed.

Remember that this can be a face-down technique, working from the front of the piece to the back. If you want the piece to dry flat under restraint after the vacuum operation, place a slightly larger sheet of plexiglass or Formica over the work before covering it with the plastic sheet. If the piece is particularly delicate, you can protect the surface from distorting under the pressure of the vacuum by placing a sheet of Ultra felt or a thin layer of foam between the work and the plastic sheet.

You can use the vacuum table for work with shallow relief shapes, including "found" objects, or casting molds, using a low-shrinkage, lightly beaten pulp such as cotton linter. Styrofoam makes a good relief surface. It is easy to model in sheet form, the water can be drawn through it, and the work can be left to dry directly on it after the vacuum process has been completed.

You can apply additional pressure to the pulp, particularly in the case of relief work, by using a sponge or large stencil brush. Be careful not to damage the plastic sheet

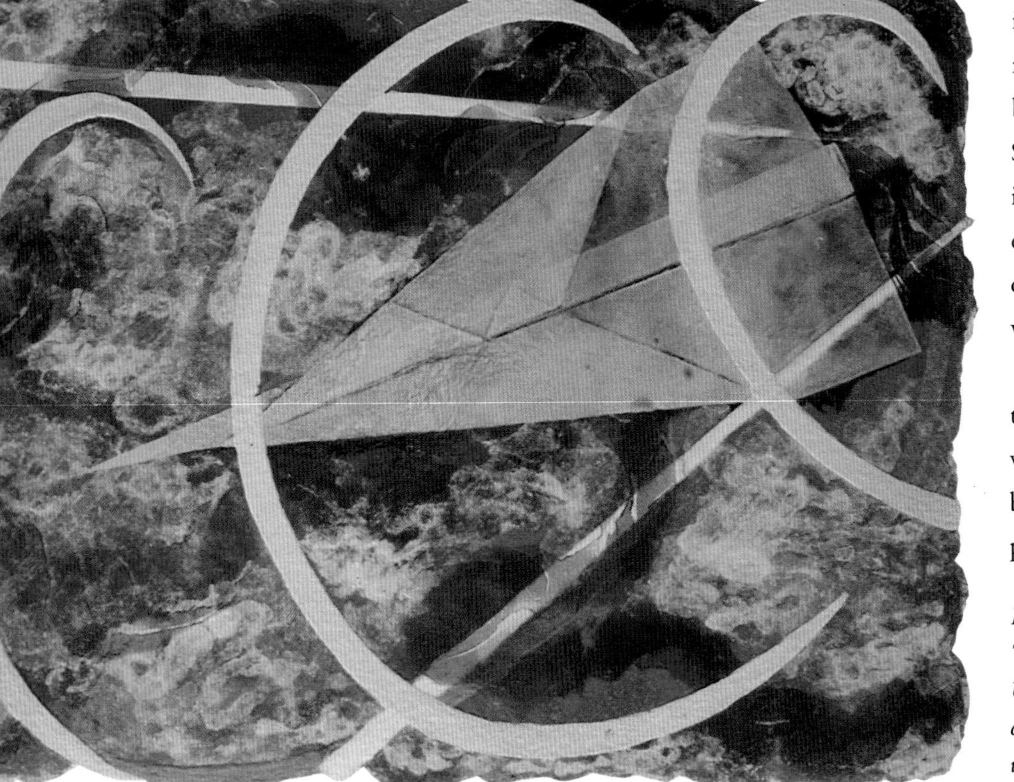

Pat Gentenaar-Torley: On the Way Up. Pulp painting created by pouring a variety of pulps and then building up layered images on a vacuum table. 39×59in. (100×150cm).

Paul Ryan: Catalan. Pulp painting. 31×20in. (79×51cm).

Gallery

Judith Faerber: Cacti. Handworked paper pulp. 24×20in. (61×51½cm).

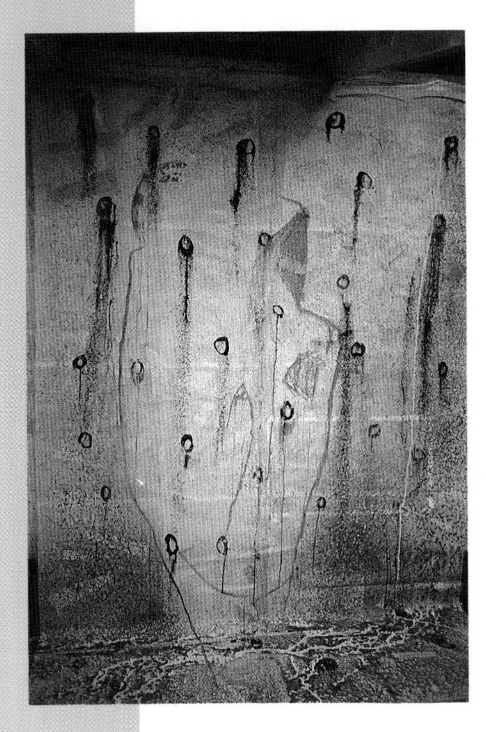

Anne Vilsboll: Dynamite. Pulp painting. Handmade kozo paper and sprayed abaca pulp. 118×98in. (300×250cm).

Laurence Barker: Untitled. Pulp painting. 27×40in. (70×101cm).

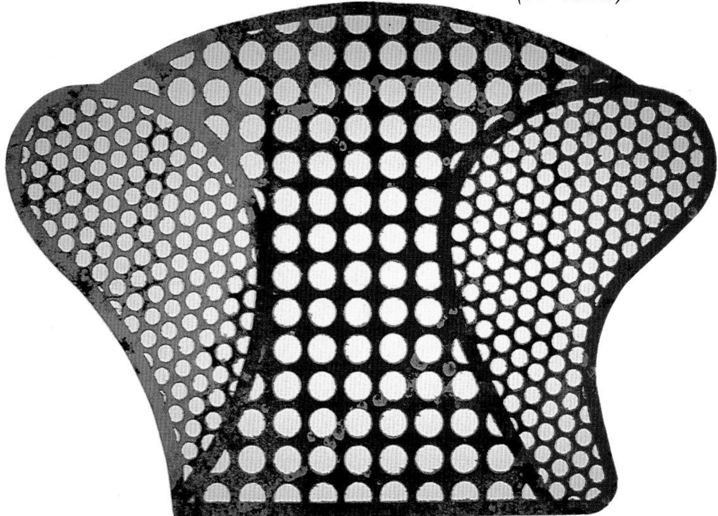

Gallery

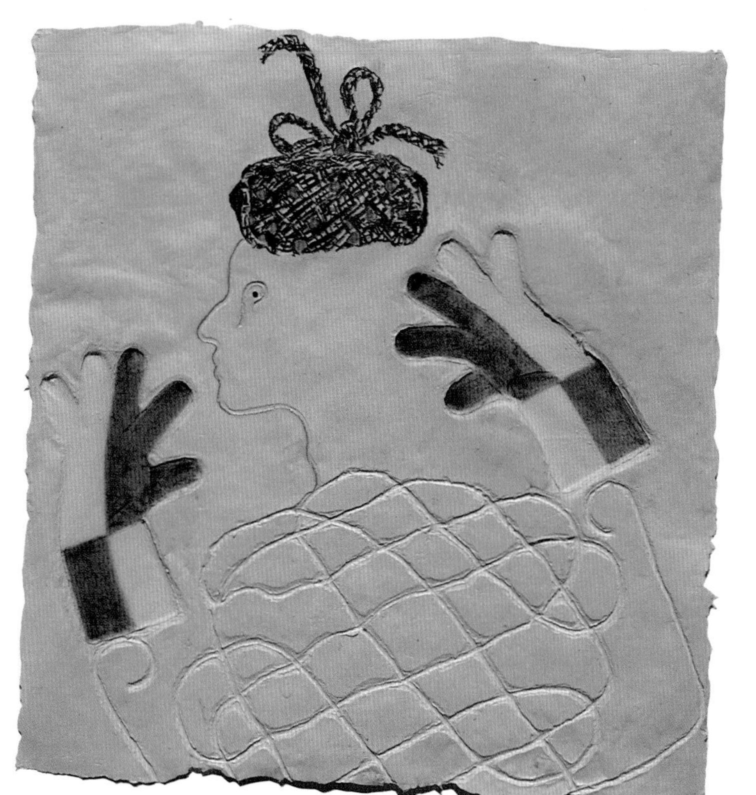

Julie Norris: Scotland. Embossed, recycled gampi tissue, pastel. 28×26in. (71×67cm).

Kyoko Ibe: And the Sea. Blind embossed handmade paper. 71×71in. (180×180cm).

Therese Weber: O.T. Handmade paper. $25\frac{1}{2}$×29in (65×75cm).

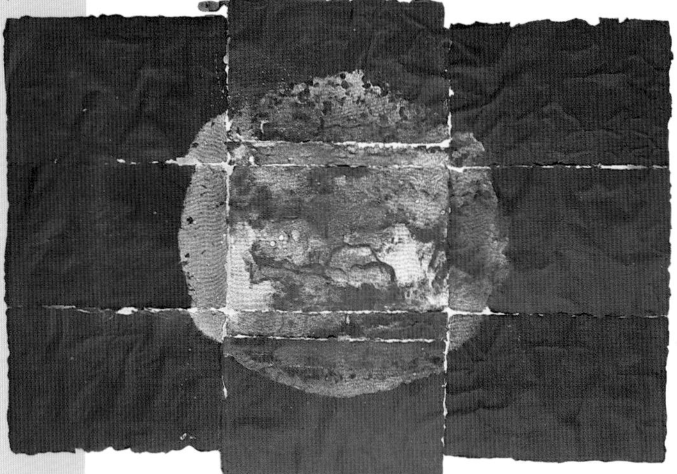

Andreas von Weizsacker: A Star is Born. Watermarked handmade paper. 61×38in. (155×96cm).

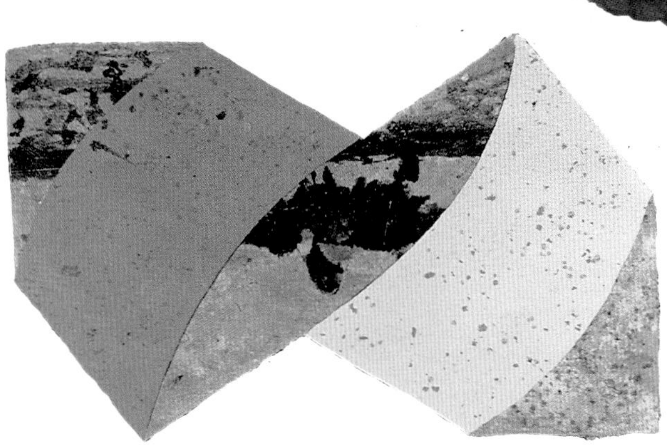

Peter Sowiski: Giff. Colored pulps, built up in shapes and overlaid laminations. 36×60in. (91×152cm).

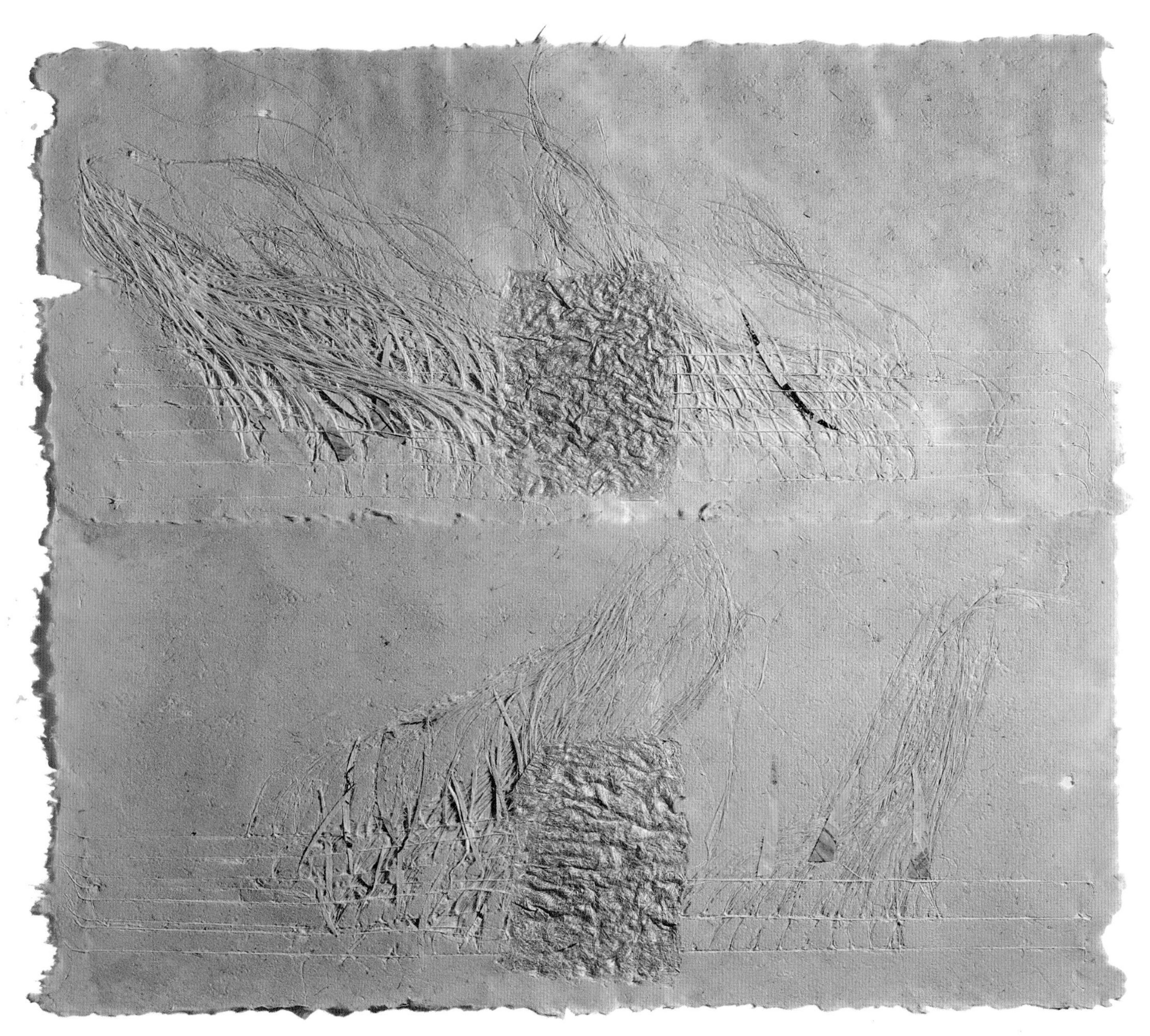

Prof. Dorothea Reese-Heim: Klangfiguren. Handmade linen and asparagus fiber paper. 35×39in. (90×100cm).

Beetroot paper made from sliced cross-sections of the vegetable, laminated by a process similar to papyrus-making.

Jean Stamsta: Bird Portrait. Recycled paper, fabric, acrylic. 52×56½in (132×143cm).

CHAPTER

4

Sculptural Techniques

Ray Tomasso:
Bronze Tile Game
(Copper Version).
100 per cent rag,
cast paper.

Jeanne Jaffe: Gardener Series: Bending Figure. Cast, hand-pigmented paper layers. 28×61×48in. (71×155×122cm).

Cast Paper

Ted Ramsay: Changing Courses (detail). Cast handmade paper, enamel, and wood. 37×32in. (94×81cm).

One of the most dramatic developments to have extended the vocabulary of papermaking is a recognition of paper's sculptural potential. Its ability to assume almost any form and still be lightweight makes paper a perfect medium for creating three-dimensional work. The cost is minimal compared with that of most sculpture materials, and the principles of casting paper are fairly straightforward. Many artists are attracted to paper for these reasons and have brought to it approaches from other crafts, such as the casting methods of sculpture or handbuilding in ceramics.

Innovative adaptations of these methods, coupled with an increased awareness of the options for preparing fibers and the handling properties of pulp, have enabled the creation of a tremendous number of intriguingly varied three-dimensional works.

Most paper sculpture is made either by pressing prepared pulp, or by layering small pieces of lightly pressed paper into a plaster or latex mold. Objects without undercuts, or areas where the paper can become trapped in the mold, can be cast using a simple one-piece plaster mold. An object "in the round" must be divided into sections for casting and the separate pieces assembled after the paper has dried. An object with undercuts can be successfully cast in a flexible rubber latex mold with a plaster back-up.

Your choice of fiber and method of pulp preparation are important. Bast fibers and seed fibers have different characteristics, which may be more appropriate for one casting technique than another. In general, it is worth noting that prolonged beating, particularly of herbaceous bast fibers (such as flax), results in a high-shrinkage paper ▶

PROJECT

Making a Plaster Mold

1 Choose a simple shape to cast from, or make your own from modeling clay. Avoid undercut areas that might hinder the removal of the mold from the cast. Model the clay directly on a Formica board. This can be used as the base of the casting box. If the object to be cast is porous, it should be waterproofed and treated with a mold release agent.

2 Make a 2-3in. (5-8cm) deep wooden frame to fit onto the base board (a simple box mould). This will determine the thickness of the mold casting. The corner joints can be screwed to allow the frame to be taken apart and reused.

YOU WILL NEED

- *Modeling clay or 3-dimensional objects (without undercut areas)*
- *good-grade plaster for mold making: U.S. Gypsum Molding Plaster No. 1 or Georgia Pacific K59*
- *mixing bucket*
- *scales (for weighing plaster)*
- *strainer (for sifting plaster) (optional)*
- *mold release agent (Vaseline, soft soap, or buffed paste wax)*
- *Formica-covered board*
- *4 lengths of wood*
- *4 brackets and/or brass screws*
- *silicone sealer*
- *water sealant or linseed oil*
- *small brush*

TECHNIQUES

- *making a mold*
- *sheetforming*
- *couching*
- *pressing*

3 Waterproof the frame, and fill any gaps between the frame and the board with a silicone sealer.

4 Pour cold water into the mold, almost to the top of the frame. Measure the water from the mold into a mixing bowl or plastic bucket with a lip for pouring. The ratio of plaster to water is critical. Follow the manufacturer's instructions very carefully. Sprinkle the pre-measured powder into the cold water. The plaster will begin to float as it displaces the water in the bucket. Wait for it to dissolve.

◀ with a greater degree of translucency than a shorter fiber pulp (such as cotton linter), or than the same fiber beaten for a shorter period. The nature of the fiber itself, the method of preparation, and the style and duration of beating can all be adjusted to suit a variety of requirements.

Shorter fibers and low-shrinkage pulps may cause fewer casting problems, but tend to require the addition of an adhesive agent, such as methylcellulose, for rigidity or strength. Longer fibers or high-shrinkage pulps may present greater shrinkage problems, but offer a wider range of casting possibilities. Shrinkage tests, based on variously prepared fibers, can provide a useful basis for developing further three-dimensional effects.

DRAMATIC VOLUMETRIC FORMS

You may have already noticed the tendency of a particular fiber to shrink or distort more than another, especially during the drying process. Sometimes the edges of a loft-dried sheet become cockled, or part of a board-dried sheet lifts off the board before the rest of the sheet has completely dried.

The forces that occur within a sheet as

5 **Gradually add the rest of the plaster. Leave for 20 seconds, or until it is saturated, then put your hand gently into the plaster and make a fist. Slowly unclench your hand and make a fist again several times. This will force out any lumps.**

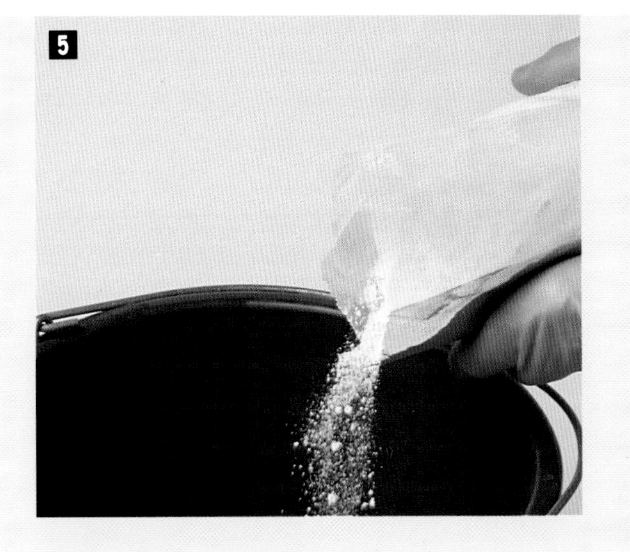

7 **Pour the plaster into the frame from the lowest point in the mold until the object is completely covered and the mold is filled. The plaster will creep over the prepared object and thus reduce the risk of trapping air bubbles. Clean off utensils immediately, as the plaster hardens and cannot be dissolved. Never pour plaster down a drain or sink, because it can set underwater. During the setting process (27-37 minutes for No. 1 U.S. Gypsum), the plaster heats up. When it has cooled and the cast is hard, detach the frame and remove the plaster mold from the model. Clean off any clay from the mold.**

6 **When the plaster has been absorbed, the mixture should resemble heavy cream. Do not stir the plaster as this will cause air bubbles and fast setting. Slow setting creates a strong mold.**

Winifred Lutz: Epiphenomenon (Phase Transition). Cast handmade flax paper and pawlonia log section. 45×120×31–50in. (114×305×79–127cm).

the hydrogen bonding between the fibers tightens are considerable. If the sheet is restrained during drying, these forces are kept in check. Most drying methods are based on this principle and are designed to minimize the undesirable effects of shrinkage. However, if the forces are allowed to resolve themselves freely, they can be used to create quite dramatic volumetric forms. Three- ▶

TECHNIQUE

Sheet Casting

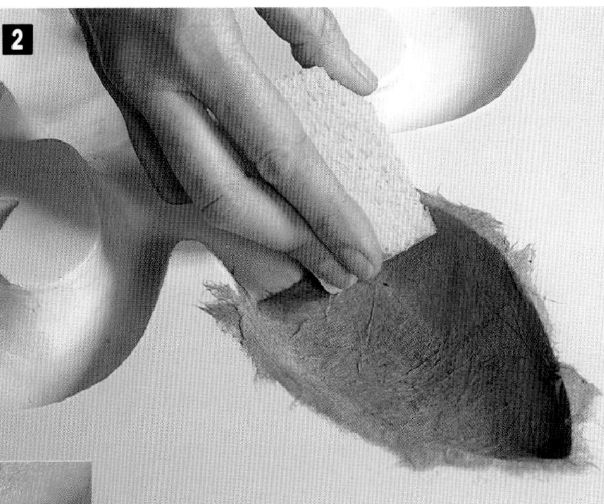

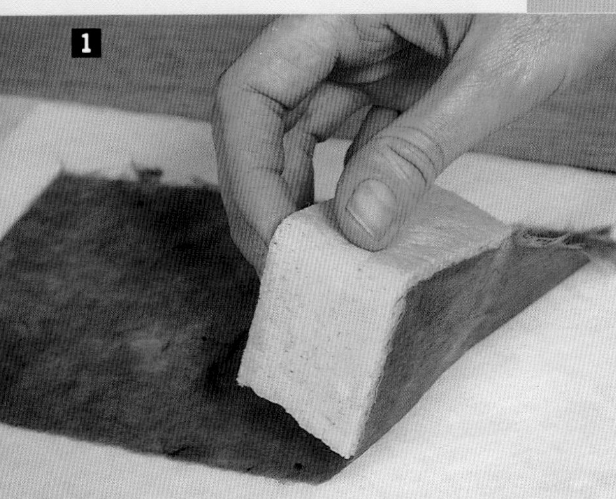

YOU WILL NEED

- *plaster mold or box mold with relief form(s) attached*
- *lightly pressed sheets of paper (low-shrinkage pulp)*
- *sponge*
- *short, stiff-bristled brush*
- *methylcellulose (or rice or wheat paste)*
- *small brush*
- *restraining form (optional)*

1 **Tear a small strip from a lightly pressed sheet. Tear against the edge of a ruler, or use a sponge to lift a section of the sheet off the felt. Feathering the edges in this way guarantees a strong, seamless bond.**

2 **Build up a layer of these pieces in the mold, making sure that the edges overlap. The pieces can be tamped into place with a short, stiff bristled brush. This will help to fuse the overlapping edges and guarantee the reproduction of surface details.**

3 **The completed first layer.**

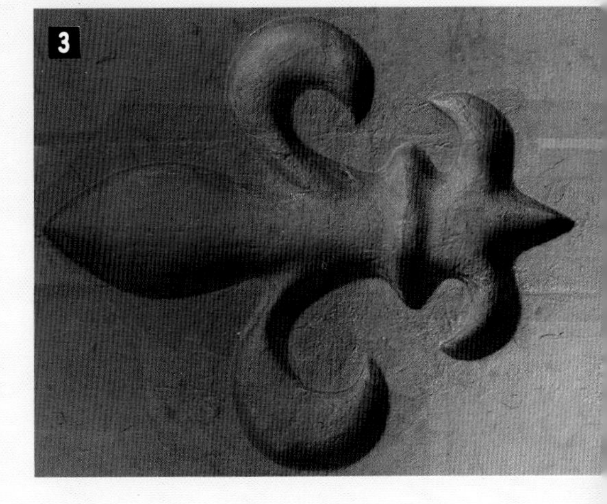

◀ dimensional effects formed during drying can be manipulated in a number of ways: by couching two sheets with different shrinkage rates against each other and allowing the pressed laminated sheet to dry without restraint; by allowing one part of a sheet to dry faster than another; or by embedding extraneous nonpaper materials between laminations of a high-shrinkage pulp to create shrinkage counterforces within a single sheet.

RECYCLED TRACING PAPER

Beating a fiber for high-shrinkage usually calls for a Hollander beater, but tracing paper offers a successful and readily available alternative to start with. Recycled tracing paper has a slippery consistency and a fairly slow drainage rate on the surface of the mold. Although its waxy, translucent appearance is lost during recycling, it exhibits a high-degree of shrinkage if allowed to dry unrestrained.

SHEET CASTING

There are two distinct methods of applying pulp to a mold or relief surface to create a paper cast: sheet, or laminate,

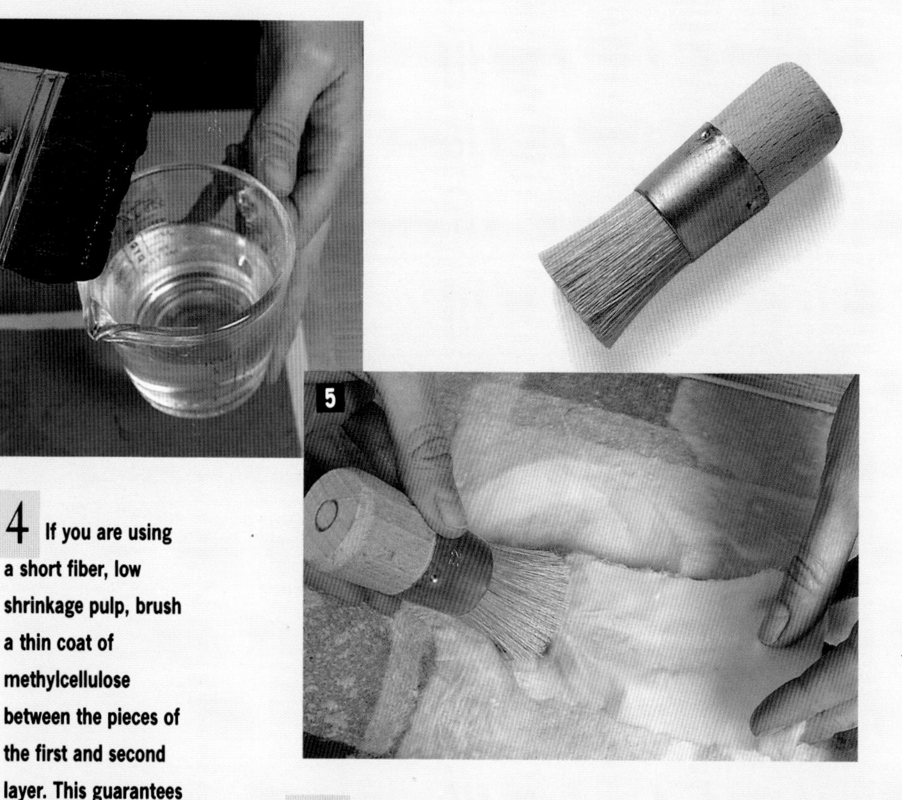

4 If you are using a short fiber, low shrinkage pulp, brush a thin coat of methylcellulose between the pieces of the first and second layer. This guarantees adhesion between laminations and increases the rigidity of the casting.

5 The edges of the second and any subsequent layers should be overlapped and placed above a seamless area in the layer beneath. Several thin layers will give a stronger cast than a single thick one. A thin first layer, using a plant fiber paper, can be backed with layers of a more readily available or economically prepared pulp. Additional layers will increase the opacity of the casting.

6 If you have used a high-shrinkage pulp, you will need to hold the paper in place as it dries. You can use a heavier weight comparatively non-shrinking paper with methylcellulose for the final layer, or a plaster back-up, or rustproof weights.

Lillian A. Bell: Under the table: fought by metaphor. Cast paper. 18×27×21in. (46×69×53cm).

casting, and pulp casting. Both methods can be used in a plaster mold, or in any other suitably prepared form.

Laminate casting is best done with freshly couched sheets of paper that have been lightly pressed. It involves layering pieces of paper in the mold and overlapping the edges where the pieces come together to achieve a seamless appearance. Sheet lamination can give a thin, ▶

7 **Leave the paper cast to dry in the mold. The plaster will absorb some of the water from the paper and hasten the drying time. Casting by this method seldom takes long to dry, but will depend on the number of layers and the humidity of the room. A fan can be used to assist the drying process.**

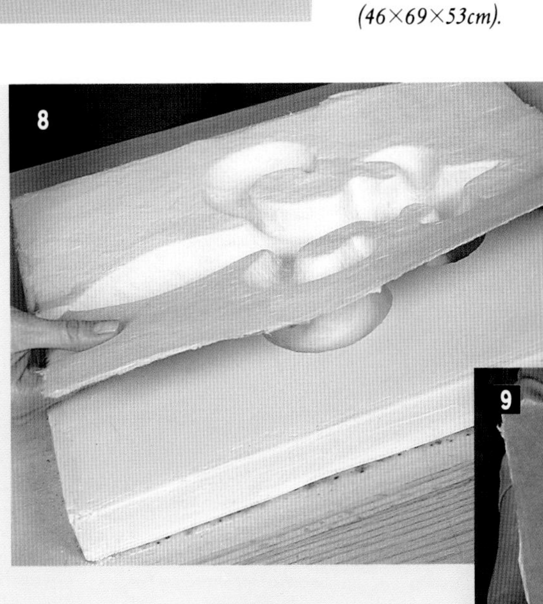

9 **Gently ease the cast from the mold until it is completely free from the plaster.**

8 **When the cast is completely dry, separate it from the mold with a sharp knife.**

light, yet strong casting with potential variation in translucency and opacity.

PULP CASTING

Pulp casting generally requires large amounts of strained pulp and will give a thicker, more opaque result than sheet, or laminate, casting. A short fiber, low-shrinkage pulp, such as cotton linters, is most commonly used. Methylcellulose, a flexible paper glue, is often added to strengthen the pulp and make the paper stiffer when dry (see p. 27). Like sheet casting, it can be accomplished using many kinds of relief surface, and there are numerous natural forms as well as manufactured objects which can be adapted to create an interesting mold. You can use the same basic box mold as for a plaster cast (see p. 101) and attach the chosen forms to the base board. Foamcore board is easily made into a simple relief surface. Some materials, such as wood, will need sealing before casting to prevent any impurities from staining the pulp as it dries.

If you cast the relief using a plain pulp, you can apply paint to the work after it has dried and been removed from the mold. Alternatively, you might com-

TECHNIQUE

Pulp Casting

YOU WILL NEED

- *box mold*
- *foamboard*
- *prepared pulp (low shrinkage)*
- *waterproof adhesive (silicone sealer)*
- *strainer*
- *sponge*
- *cheesecloth or muslin (optional)*
- *methylcellulose*

TECHNIQUES

- *making a box mold*

1 **Cut out the elements that make up the relief.**

2 **Assemble the relief, using a waterproof glue.**

3 **Position the relief in the casting frame.**

Nancy Thayer: Temple View. Cast paper and acrylic paint. 57×43in. (145×109cm).

bine different colored pulps and include various embedments during the final stages of the casting process.

Pulp casting does not usually confer the same bonding strength as sheet or laminate casting. Although the pressure exerted when sponging the pulp in the mold compresses the pulp, it may be necessary to add a small amount of methylcellulose to the pulp beforehand.

4 **Strain the pulp to an applesauce consistency. Pulp casting requires a lot of pulp. It is better to overestimate the amount you need than to run out halfway through.**

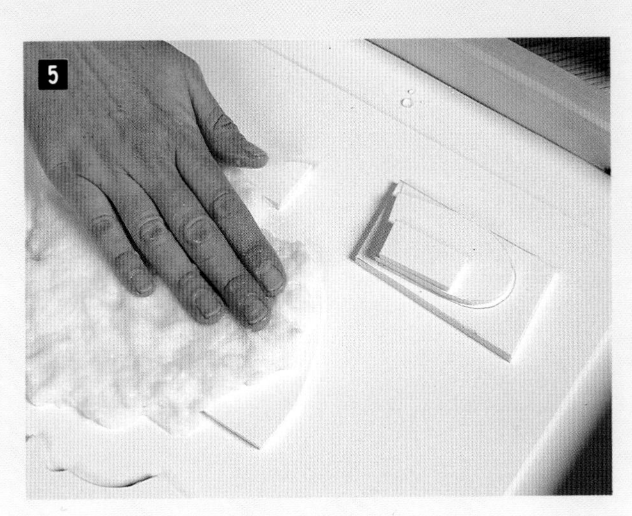

5 **Scoop a handful of strained pulp into the mold and pat it into place.**

6 **Continue to fill the mold with the pulp, patting it in place around the edges of the relief. Overlap the edges of the scooped pulp, and layer several handsful until you have built up a thickness of 1-1½in. (25-37mm). Any air that becomes trapped during this process can be released by piercing the layers with a pointed tool.**

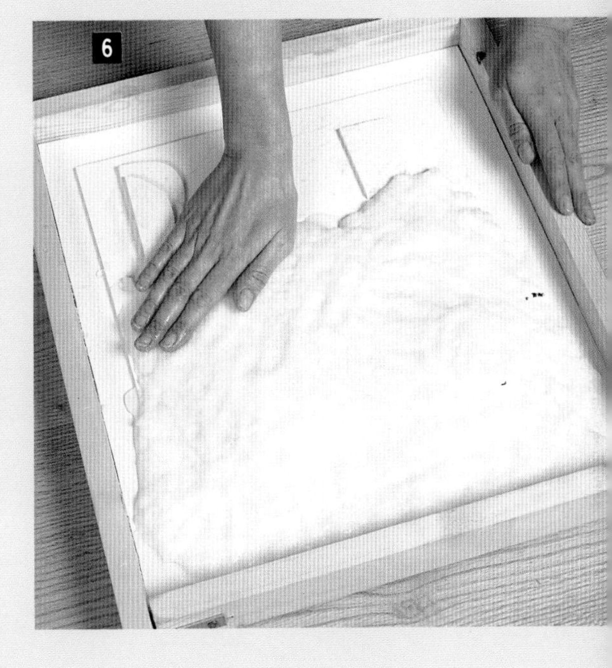

7 **Slight sponging between handsful will help to create an even, well-bonded layer. If you need to reinforce the pulp, soak some pieces of cheesecloth or muslin in methylcellulose and lay them between half-inch (12mm) layers of partially sponged pulp, making a sandwich of pulp, reinforcement, pulp.**

8 Gently sponge the excess water from the pulp, working from the center or highest points of the relief and moving out toward the edges.

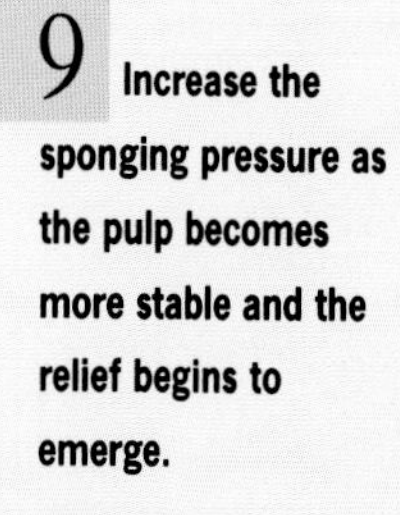

9 Increase the sponging pressure as the pulp becomes more stable and the relief begins to emerge.

10 Continue until you have removed as much water as possible. The pulp will compact down to at least half its original thickness.

11 A half- to one-inch (12-25mm) casting can take several weeks to dry. Forced or uneven drying can cause shrinking and warping, so it is best to leave the relief to dry naturally without the use of fans or heaters. Remove the frame and, when the piece is completely dry, lift the cast paper relief off the mold. You can keep the foamcore board as part of the piece or remove it as desired.

PROJECT

Volumetric Forms

YOU WILL NEED

- *prepared high-shrinkage pulp, i.e. recycled tracing paper*
- *inclusions: thin brass rods (from model-making shop)*
- *mold and deckle*
- *couching cloths*
- *pressing boards*

TECHNIQUES

- *sheetforming*
- *couching*
- *embedding*
- *laminating*
- *pressing*

1 Couch a freshly formed sheet and arrange several thin brass rods across it. You can experiment with the number and spacing of the rods and with different materials, such as strips of plastic, thin wooden sticks, or a grid of cotton or linen thread. The degree of rigidity or flexibility of the nonpaper inclusions will affect the displacement that occurs within the sheet as it dries.

2 Couch a second sheet on top of the rods and base sheet. After pressing, remove the sheet from the felt, and place it on a clean surface to dry. The sheet must be allowed to dry without being restrained, so do not brush or roller it onto the drying surface.

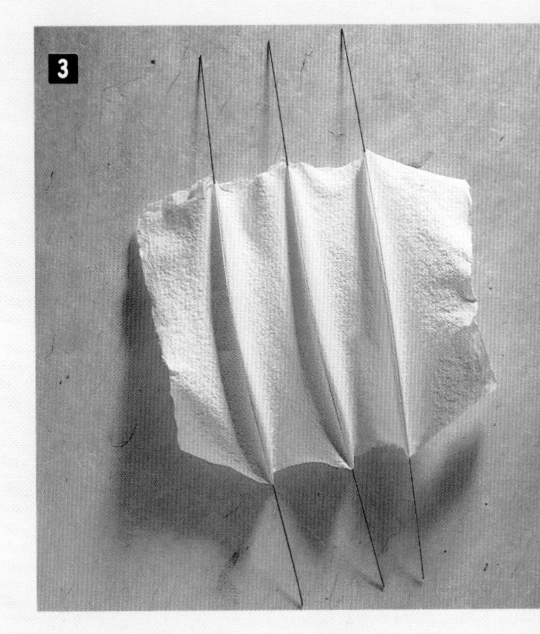

3 The paper will shrink as it dries, but the rods will not, and the sheet will therefore contract between them. Different shapes will be produced according to the degree of shrinkage of the paper, the resilience of the inclusions, and the way in which they cross the sheet.

Jody Williams: Escaping Depressions. Cast paper. 15×20in. (38×51cm).

Gallery

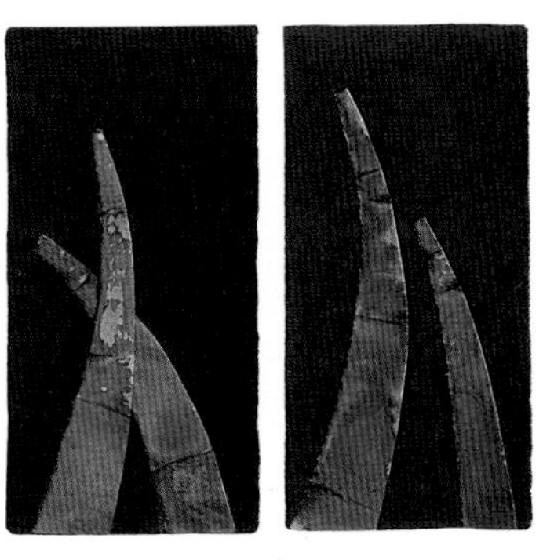

Cathrine Schei: Cut IV and V. Cast handmade paper. 12×24in. (30×60cm) each.

Lilian A. Bell: Revolving Door Policy. Cast paper, mixed media. 59×44×26in. (150×112×66cm).

Ted Ramsay: Life Strategy II. Cast paper, enamel, wood. 46×46in. (117×117cm).

Josephine Tabbert: Papierarchitektur (detail). Handmade paper. 6×16in. (16×40cm).

Jeanne Jaffe: Totem Series – The Three Muses. Cast handmade paper, oilstick, gouache. 82×84×18in. (208×213×46cm).

Donna Koretsky: Around These Lines. Unbleached flax fiber applied to collapsible latex molds. 7×84×7in. (18×213×18cm).

Helmut Becker: Tree Skins: Totemic Tower, Totemic Burls, Totemic Roots. Spray cast fiber flax paper pulp.

CHAPTER

5

Contemporary Adaptations

Guy Houdin: Patak pense à Paul. Woven strips of painted and twisted brown (kraft) paper.

Vito Capone: Libro-Sovrapposizioni. Book object made from "carved" cotton paper. 8×9×5in. (20×24×12cm).

Paper and Books

A collection of books with varied shapes, sizes, bindings, and decorative covers.

Long before the development of paper, when various systems of writing began to replace the oral traditions and memorized texts of early peoples, "books" were used to record important events, customs, codes of law, religious myths, and rituals.

"BOOKS" OF CLAY

Clay tablets and rolls of papyrus are the earliest extant prototypes for the book as we recognize it. Clay tablets of various shapes and sizes were the main means of communication in Mesopotamia for nearly 3,000 years. The clay was inscribed while it was damp and soft, and either dried in the sun or kiln-baked to make it more durable. Although it was not possible to bind the tablets together, "books" were made up of several tablets, each named, numbered, and carefully indexed. Clay "envelopes" were often fashioned around the tablets to guarantee their preservation.

BUTTERFLY BOOKS

In China, the earliest bound books were thin slips of bamboo or wood joined by cords. These gave way to silk (and later paper) scrolls. The inconvenience of unrolling a scroll to read a particular section was overcome in a number of ways. One of the simplest was to fold the scroll like an accordion, attaching covers to the first and last sections. Sometimes the first and last pages were bound to a single cover, leaving the folded pages free to flutter out from the binding (a "flutter" or "whirlwind" book). Later, to alleviate the stress on the folds of such books, single sheets of text were folded face inward, stacked for binding, and pasted along the folded edges so that the pages stood out like the wings of a but- ▶

PROJECT

Single-section Book

YOU WILL NEED

- *a selection of papers (preferably handmade): not too thick for the inside pages; slightly heavier weight for cover*
- *pencil*
- *ruler*
- *steel square*
- *cutting board*
- *scalpel or craft knife*
- *short-eyed needle (large enough for thread)*
- *heavy thread (preferably linen)*
- *awl*
- *scissors*
- *bone folder*
- *binder clip*

1 Take the sheets for the inside pages, and fold each one in half, with the grain direction parallel to the fold (see p. 120). Use a bone folder to score the fold lightly.

2 Slip the folded pages inside one another to form a section. The top edge is the "head," and the bottom edge is the "tail." The open edge opposite the folded spine is the "fore edge."

3 The measurements for the cover should allow a 1/8in. (3mm) margin beyond the page edges at the head and tail, and 5/16in. (8mm) at the fore edge. This allows for the take-up when the cover paper is folded around the section, giving an equal 1/8in. (3mm) margin on each edge. If the section pages are the same size as the cover sheet, trim them down slightly before you fold them. If you want to keep the deckle edges of the pages, however, you will need to choose a larger cover paper, and cut it down to give the marginal overlap described above. Use a steel square to make sure you get right-angled corners, and a ruler to measure straight parallel edges between the head and tail, the spine and fore edge. Remember to check the grain direction before you cut the paper.

4 Fold the cover paper in half along the grain, and score the fold with the aid of the bone folder. Do not press too hard, or you might stretch or mark the paper. A light but firm pressure along the length of the fold is all that is needed. Slip the section inside the cover, leaving an equal margin at the head and tail.

◀ terfly when each one was turned (hence a "butterfly" book). This marked the beginning of the shift away from the scroll and concertina format toward the book shape exemplified by the parchment or vellum manuscript volume.

JAPANESE BOOKS

The essentials of Japanese book production were inherited from China. Scrolls, accordion or concertina books, album, and "butterfly" books were all based on Chinese models; but during the 11th and 12th centuries, the development of a multisection stitched binding (*retchoso*), curiously similar to Western binding of folded sections, was unique to Japan. Several sheets of paper were stacked and folded in half to form a section and a number of sections were stitched together through the central folds.

POUCH BINDING

Today, characteristically Oriental-style books are produced by pouch-binding, which replaced other methods in China and Japan during the Edo period (1603–1868). The pouch-bound book (*fukuro toji*) consists of sheets that are folded in half and stitched through the edge opposite the fold. Each page forms, in effect, a pouch that is open at top and bottom. The text is, therefore, written or printed

5 **Open out the book to the inside ("gutter") fold. Find the mid-point of the gutter fold, and mark it lightly with a pencil.**

6 **Mark two holes on each side of the mid-point, 1in. (25mm) from the top and bottom edges of the section. Then, making sure that the pages are still in position, prick three holes straight through the pages and the cover at these points, using the awl.**

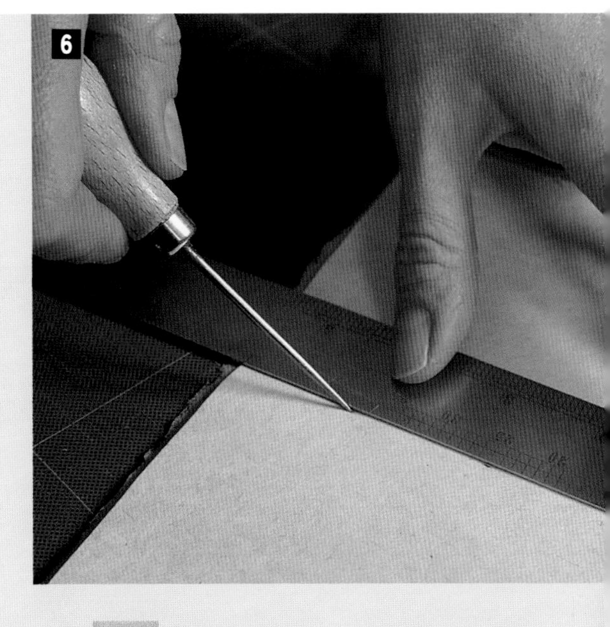

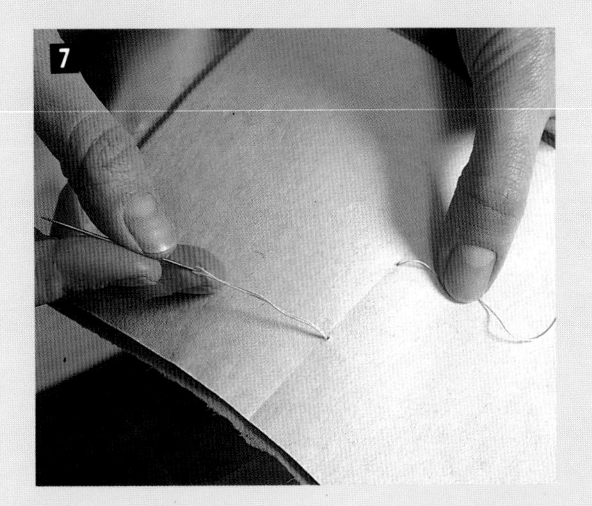

7 **Cut a length of thread (approximately two-and-a-half times the length of the section) and thread the needle. Holding the cover and inside section together so that the pages do not slip out of position, start sewing from the inside of the book by passing the needle through the center hole. Hold the tail of the thread with your thumb against the middle of the book. Return the needle from the outside through the top or bottom hole.**

Kathy Crump: Bald Eagles in Ventana. An oriental style binding with handmade paper covers of philodendron fiber over abaca (left) and sample pages (below).

on one side only. A sheet of thin cardboard can be inserted into the "pouch" to prevent water-based ink from bleeding through to the other side of the folded page. The most common type of pouch-bound book has a straightforward sewn binding with stitching holes at four points along the spine, but more elaborate stitching patterns, sometimes with decorative inserts along the spine, are also used. Unlike traditional Western ▶

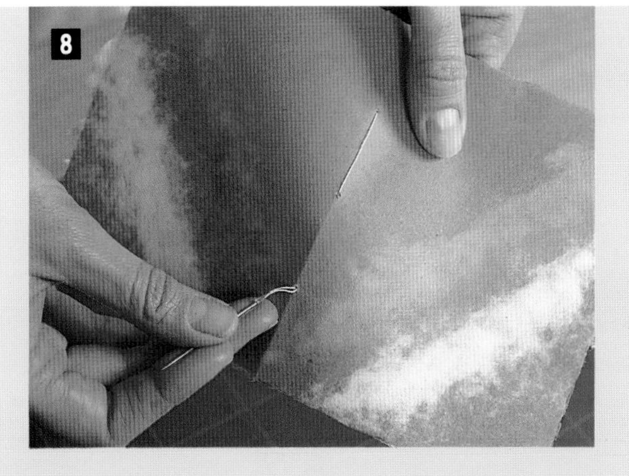

8 Pass over the center hole and take the needle out through the other hole.

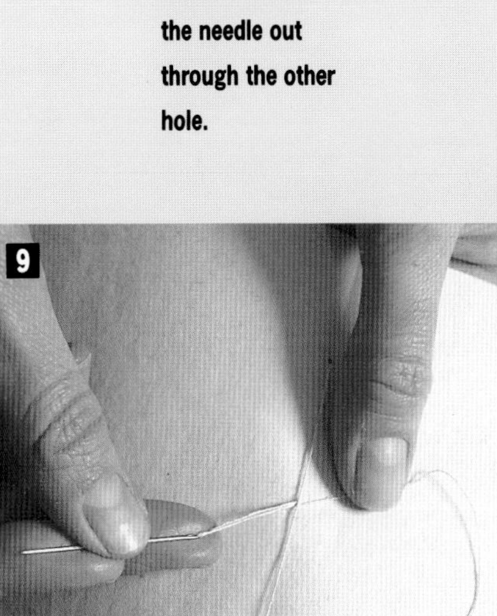

9 Return the needle through the center hole so that the thread comes up on the opposite side of the long center stitch from the tail of thread you have been holding with your thumb.

10 Pull in the thread on each side of the long center stitch, and tie it securely using a square knot. Cut off the loose ends to about ½in. (12mm).

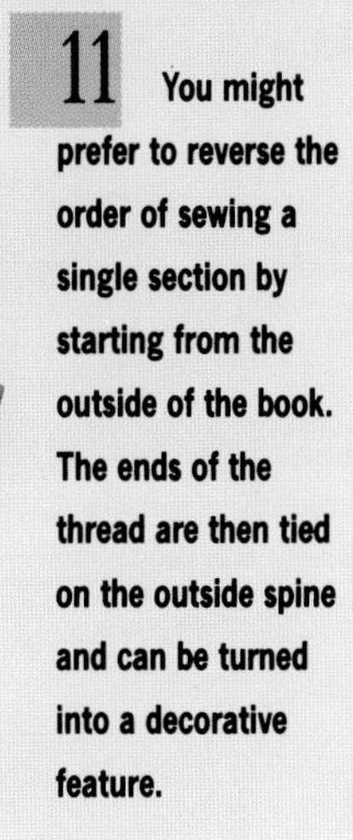

11 You might prefer to reverse the order of sewing a single section by starting from the outside of the book. The ends of the thread are then tied on the outside spine and can be turned into a decorative feature.

books, the Japanese pouch-bound book usually has a soft cover.

THE BOOK AS OBJECT

The book has always been a powerful object, revered for its ability to disseminate knowledge and proclaimed as a symbol of truth. The book has also been feared and reviled, and many a book has been burned, drowned, censored, or suppressed. The outward appearance of

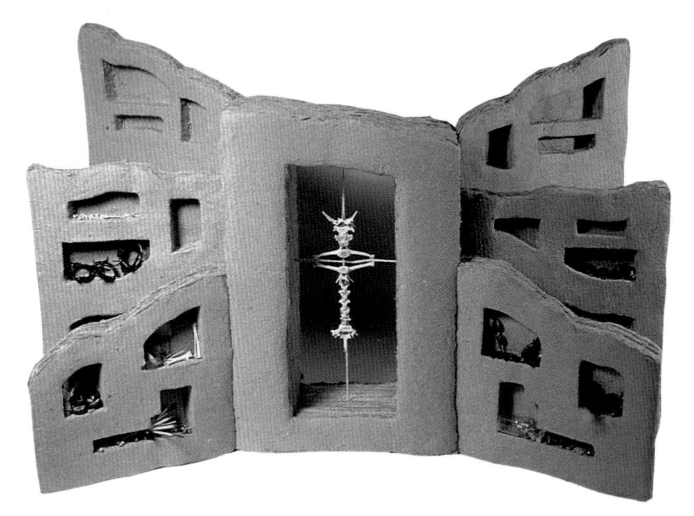

Dianne L. Reeves: Cause for Interpretive Dissension (front and back views). Handmade paper of dyed sisal, bones, scales, and pods. 25 × 12½ × 15½in. (63 × 32 × 40cm).

PROJECT

Japanese Pouch Binding

2 **Make sure that the folded edges are aligned by knocking them down against a flat surface.**

1 **Fold each page in half, with the grain direction parallel to the fold. Score the fold lightly, using a bone folder. Then assemble the sheets with the folded and open edges corresponding.**

3 **To trim the deckle edges, place the block of folded sheets on a flat cutting surface, and use a ruler and sharp cutting knife to trim the head and tail at right angles to the folded fore edge. You may find it helpful to place a weight on top of the pages to prevent them from slipping while you trim them and mark them for binding.**

the book has altered over the centuries, but its basic format has long remained unchanged.

In recent times, however, artists have reexamined and extended the definition of a book. Where the visual and tactile qualities of the book override its content, the book has become a work of sculpture. It can be a game, a toy, a conceptual vehicle; the typography and inner dressing of the book can be distorted to a degree of illegibility – it can remain a "closed" book; or techniques of paper engineering can transform it into a pop-up construction or performance piece; there can be a unity between the meaning of a word, its typographical form, and the shape of the book – or the cover can be used to belie the contents. The familiar space of an open book becomes the stage for the dynamics and disclosures of a performance. ▶

4 **Hold the knife vertically against the ruler and cut through one or two pages at a time. Use a steel square to check that all the corners are right angles, and measure the distance between the spine and fore edge to make sure they are parallel before you trim the spine.**

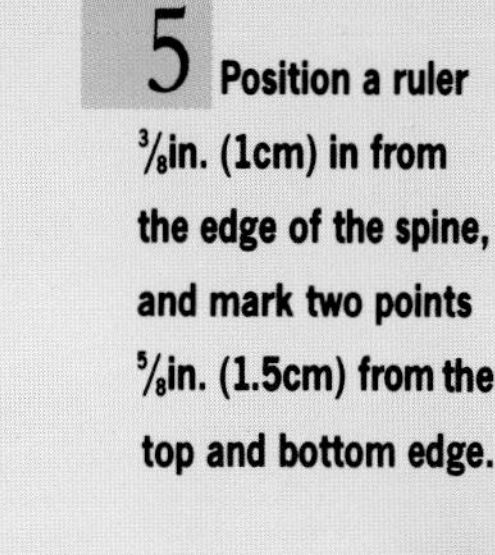

5 **Position a ruler 3/8in. (1cm) in from the edge of the spine, and mark two points 5/8in. (1.5cm) from the top and bottom edge.**

6 **Measure the distance between these two points into three equal sections, and mark up the two further dividing points. Use an awl to pierce four holes straight through the block of pages at the marked points.**

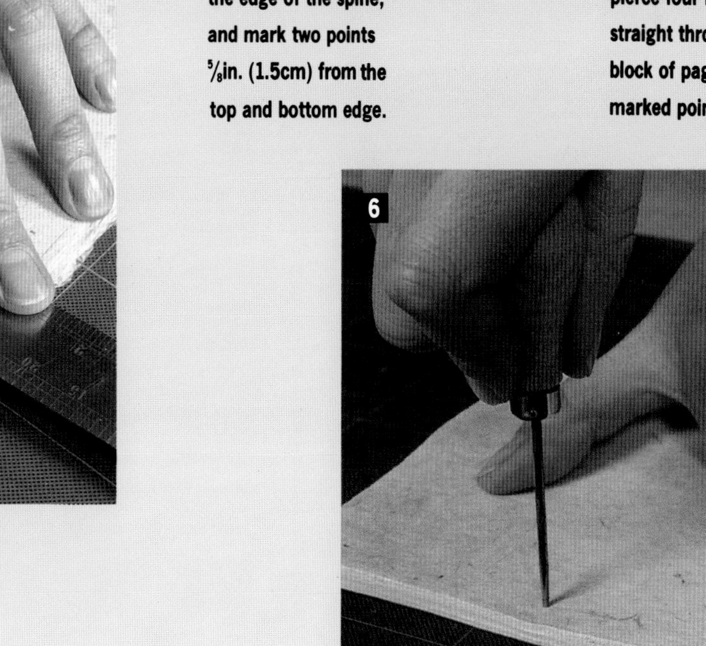

◀ MAKE YOUR OWN BOOK

The techniques involved in making a book and the skills required to decorate it and to design and print the text and illustrations were fully developed by the 16th century. Each element of the traditional handmade book was, however, a specialist trade, and sometimes even a protected one. It was only with the emergence of private presses as part of the Arts and Crafts Movement in England during the late 19th century that the tasks of papermaking, printing, and binding were combined. Nowadays new techniques and the expansion of printing technology in areas such as photocopying, computerized graphics, and quick binding mean that many elements of the handmade book are no longer "trade secrets," and that anyone can make a book – out of almost any material.

This section presents two simple but effective methods of making a book, for which the minimum of sewing and binding equipment is needed. You may use papers of any kind, as long as they are not too thick. The format of your book will be determined by the shape, size, and grain of the paper you use.

FINDING THE GRAIN

Before you begin any binding operation – even if you are only folding a sheet to

7 **Cut two pieces of paper for the front and back cover and fold them in half. (You can fold and trim them with the pages, so that they match exactly.) The grain direction of the cover and inside pages must correspond. Use one of the pages as a template, align the edges, and mark corresponding sewing holes in both folded cover sheets. The holes in pages and cover should be large enough for your needle and thread to pass through three times.**

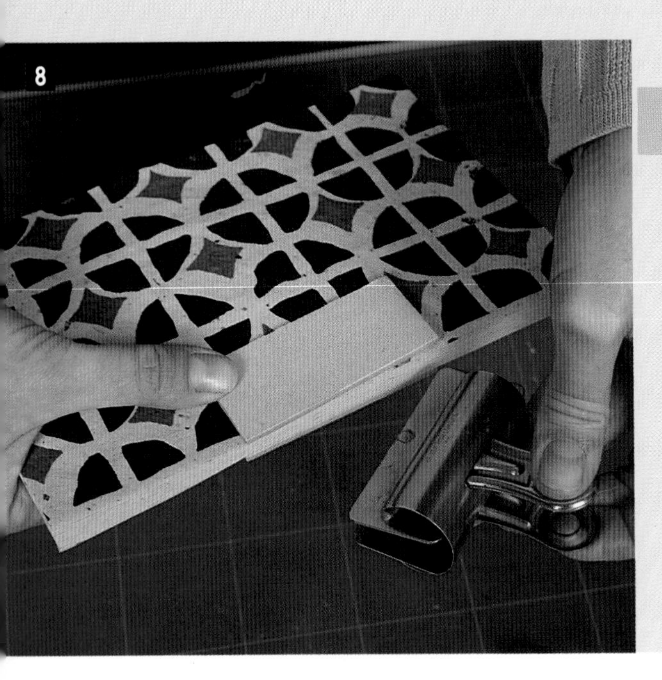

8 **Position the covers on both sides of the block of sheets. Place two small pieces of cardboard against the middle of either fore edge, and secure them with a strong clip. This will prevent the pages and cover from slipping out of place while you are sewing.**

9 **If you want to make a special feature of the binding, use colored silk or embroidery thread to sew the book together. It should be about three-and-a-half times the length of the book. Start sewing from one end of the book, using a running stitch to reach the opposite end. Hold the tail of the thread with one hand against the side of the book as you sew.**

try out different page sizes, or wrapping a cover experimentally around a book – you must establish the grain direction of the paper.

The grain direction refers to the orientation of fibers in the paper, much like the grain in a piece of wood. Although it is widely believed that Western handmade paper has no grain direction, it is virtually impossible to form a sheet in which the fibers are aligned equally in all directions. The grain direction can be determined by one or more of the following indices:

- A sheet of paper will tear more readily along the grain than against it.
- A sheet of paper will fold more easily in line with the grain than across it.
- Paper shrinks and expands more in one direction than another and, when moistened on one side, will curl parallel to the direction of the grain.

Ignoring the grain in paper can result in badly registered printing, or a book that will not open properly. The grain of all the materials in both simple and more complex binding operations must run in the same direction and should be parallel to the spine of the book.

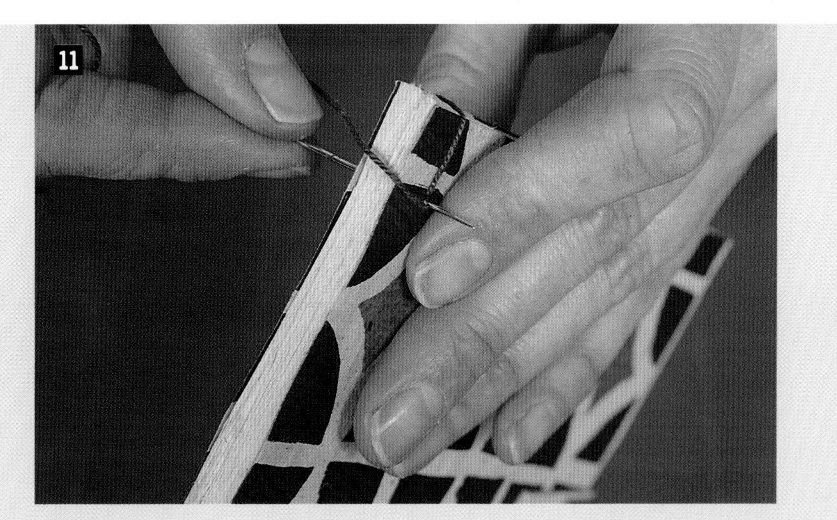

10 **When you reach the last hole, take the thread around the end of the book and return the needle through the same hole.**

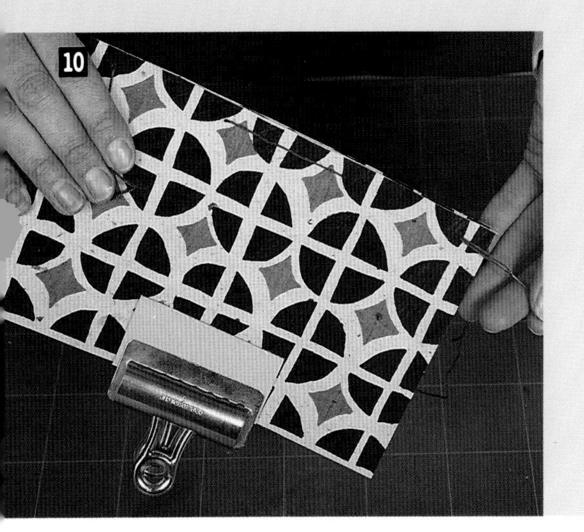

11 **Then make a stitch around the spine and come back through the same hole once again. Sew back along the length of the book, this time passing the thread around the spine at right angles to the running stitch at each hole. You should finish through the same hole as you started from.**

12 **Tie the two loose thread ends into a neat square knot close to the last hole. Then take the needle down through the hole so that the knot disappears into the spine. Cut off the ends of the thread as close to the hole as possible.**

Obi sash woven by Sadako Sakurai who is one of the few remaining Shiroishi Shifu weavers in Japan. Once spun, the paper thread is dyed and then woven with either a silk or cotton warp.

Paper and Textiles

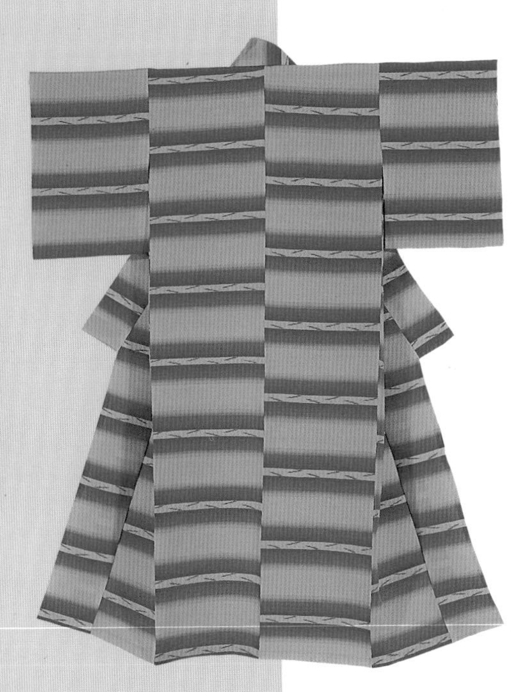

Kinujifu kimono by Sadako Sakurai. It takes her several weeks to make enough material for a single kimono.

In Japan, one of the most distinguished uses of paper is in the making of fabric for clothes. The first paper garments were made by Buddhist priests in the 10th century and called *kamiko* (from the words *kami*, paper, and *koromo*, a priest's robe). Kamiko robes are still made and worn by Buddhist monks for ceremonial purposes at the Todai-ji Temple in Nara.

Kamiko is made out of kozo paper, which is treated with a vegetable starch called *konnyaku*. The process renders the paper strong, pliable, and waterproof. The sheets are glued together to form lengths of paper fabric which can then be cut and stitched. Kamiko was made into outer garments, such as coats and jackets, as well as vests and underwear. It was a popular form of clothing among the poor and was greatly valued for its warmth in winter. The wealthier nobility wore dyed, stenciled, and embroidered kamiko of great elegance for both everyday and ceremonial wear. Household items, such as pillows and blankets, were also made out of kamiko.

WOVEN PAPER FABRIC

In contrast to non-woven kamiko, the paper fabric known as *shifu* (from *shi*, meaning paper, and *fu*, meaning woven cloth) is made by weaving twisted strips of paper. The craft of weaving with paper was first recorded in Japan in 1638. It originated among peasant farmers, who made paper during the winter, and skilled weavers, who developed a technique for manufacturing paper thread. The thread was dyed and used as the weft, with cotton the most common choice for the warp. Other forms of shifu include *kinujifu*, made with a paper weft and silk warp; *asajifu*, which has a paper weft and linen warp; *morojifu*, woven ▶

PROJECT

Shifu

YOU WILL NEED

- *handmade or collected papers (as recommended)*
- *pencil*
- *ruler*
- *cutting board*
- *scalpel or craft knife*
- *cotton towel*
- *bobbin winder (or spinning wheel)*

1 Fold a piece of paper in half with the grain of the paper perpendicular to the folded edge. (To find the direction of the grain, tear a test sheet two ways, down one side and across the other. The paper will tear more easily along the grain.)

2 Fold back the two ends in opposite directions so that they overlap the first folded edge by about ¾in. (2cm) and form a W-shape. Several sheets of paper can be folded together in this way.

3 Place the folded sheet of paper on a cutting board, and measure equal widths along the top and bottom edges. A quarter-inch (6mm) width is best to start with.

4 Cut from the inside fold (top) to the two folds (bottom), taking care to cut just through but not beyond the inside fold (leaving the ¾in. (2cm) selvage uncut). Unfold the paper and check for any uncut parts.

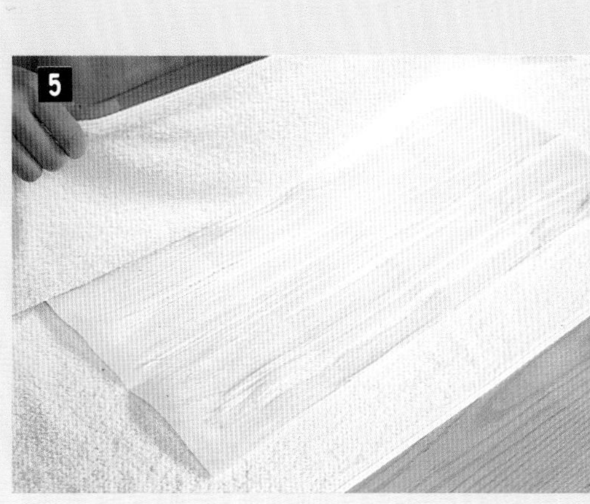

5 Open the paper out and lay it on a dampened towel. Fold the towel over the paper and place it in a clean plastic bag. Leave the paper for 7-8 hours or overnight. Several sheets can be dampened together.

Shifu hand woven by textile designer Sally Anne Watt using mercerized cotton and silk warp and plyed paper weft thread.

with both a paper warp and weft; and *chirimenjifu*, a crêpe-like fabric.

During the Edo period the wives of farmers and fishermen recycled kozo paper from used account books to make shifu working clothes. A finer cloth, considered as valuable as silk, was made in Shiroishi City by samurai family members for the Japanese nobility and given as presents to high-ranking officials. Traditional shifu garments include the kimono, haori (a short jacket), and obi (the wide sash worn over the kimono). Twisted paper thread was also woven for handkerchiefs, tablecloths, and even mosquito netting.

The technique for making shifu thread is simple and ingenious, but requires a certain amount of perseverance and practice. Almost any kind of paper may be used, although some are resistant to the twisting process and others will not be strong enough to withstand it. Handmade Japanese shifu

6 Remove the softened paper carefully, holding the sheet by the uncut ends, and place it on a porous cement block or other slightly rough surface, such as carpet backing. Traditionally, round river rocks or *tatami* mats provided the rolling surface. Using the palms of both hands, gently roll the paper strips across the surface and back again. Repeat this five times.

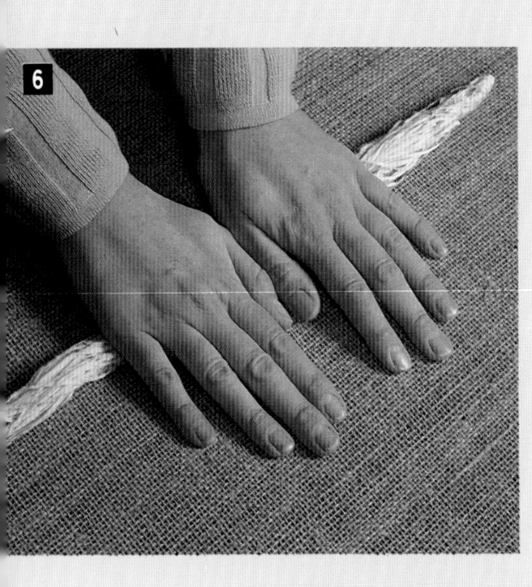

7 Remember to shake the sheet every so often during the rolling process to straighten the strips. Lift the sheet and separate any tangles with a sharp tug so that the strips are parallel.

8 Then put the sheet back and start rolling again, gradually increasing the pressure until the paper begins to twist. This may take about ten minutes and must be done quickly before the paper dries out.

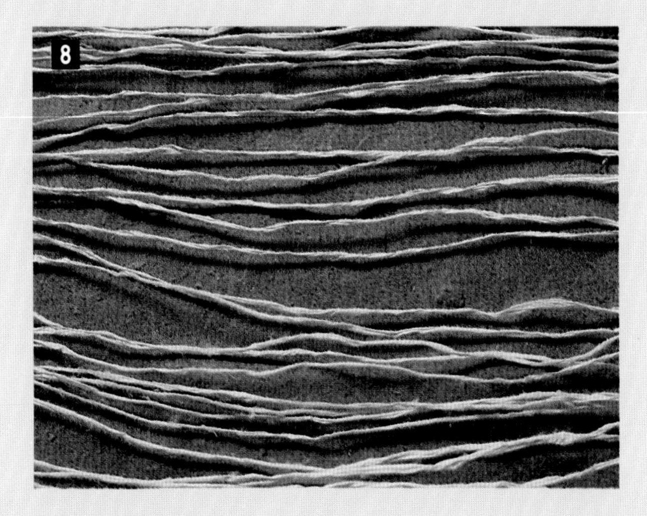

paper provides a particularly strong paper thread because the mold is shaken only from front to back during sheet-forming, thus aligning the fibers in a single direction. This means the paper has the structural strength it needs when it is cut into fine strips and pre-spun.

Begin with your own paper, if possible made by the Japanese sheetforming method. The best papers to use are Japanese handmade papers. They should be free of any imperfections: tiny bits of bark make it more difficult to cut the paper. Alternatively, experiment with a selection of different papers, such as thin gift wrapping paper, computer paper, or dress pattern paper. Each will handle differently and produce an interesting variation of pattern and texture. Printed papers produce an especially interesting "ikat" effect when woven with a plain warp thread.

In most cases, shifu fabric is woven with a cotton or silk warp thread and a paper weft thread. To weave colored stripes or patterns into shifu cloth on a simple hand loom, you can either dye the paper thread, or use a colored paper to make the thread.

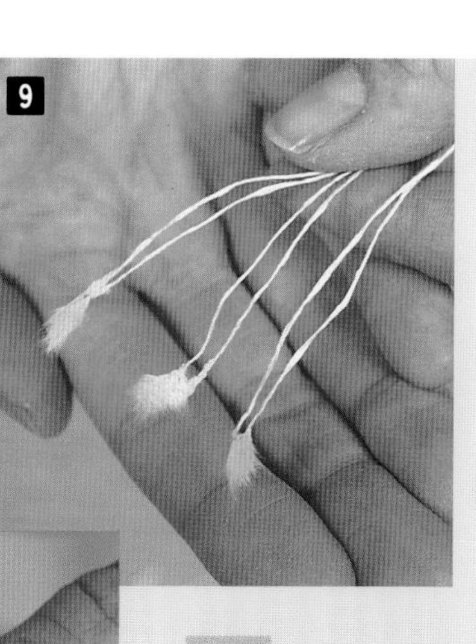

9 **Tear through the uncut edges alternately on the top and bottom selvage to create a continuous thread.**

10 **The remaining untorn joints, or "seeds," can be torn down to $\frac{3}{8}$in. (1cm) and twisted back on themselves.**

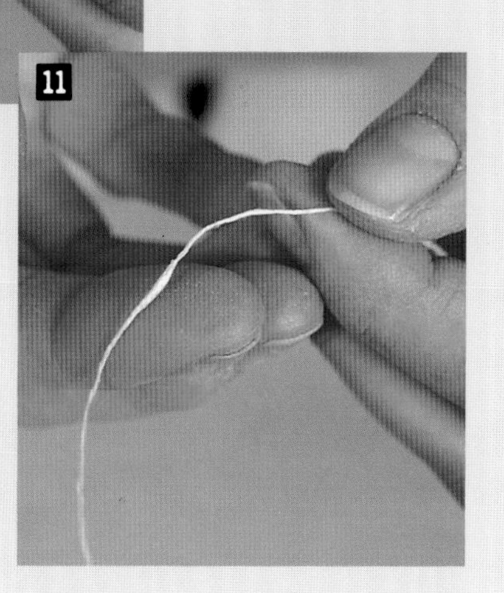

11 **The thread will be slightly thicker at each of these points, and this is a characteristic feature of shifu fabric.**

12 **Spinning the paper increases its strength and can be used to join threads from separate sheets. Attach the end of the thread to the bobbin or paper quill on the winder. Begin spinning by turning the wheel slowly with one hand and holding the paper thread in the other hand between your index and middle fingers, supporting it with your thumb.**

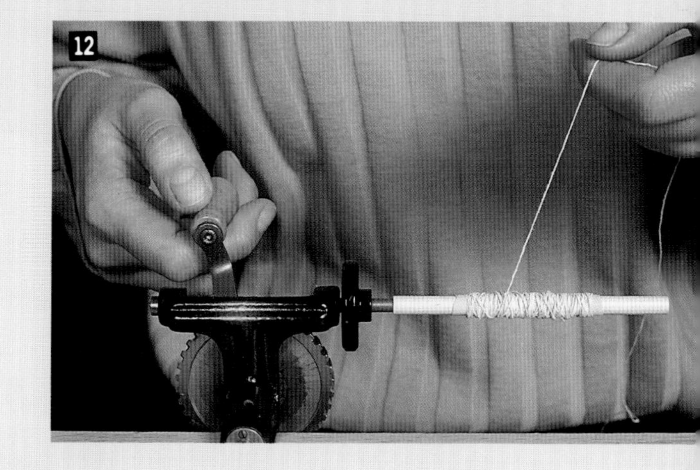

13 **Extend your arm away from the winder along the thread as you turn the wheel. When the spun paper is an arm's length from the winder, wind it steadily onto the bobbin. Raise the thread vertically as you reel it onto the bobbin winder to prevent it from untwisting. Then spin another arm's length of thread as before. Once spun, steam the thread on the spool for 20 minutes to stabilize the twist.**

Helena Sellergren: Mandala of Light. Paper, metal net, and metal. 31½×31½in. (80×80cm).

Paper and Light

Japanese handmade paper is remarkable for its combination of lightness, translucency, and exceptional strength. These qualities have been an important element in the construction of traditional Japanese houses. Sliding *shoji* doors and windows are made with paper and allow a diffused light to enter the room. *Fusama* (room dividers) are also covered with decorative paper such as *uchigumo* (cloud paper).

Lightness and translucency make paper an obvious choice for lampshades, screens, and blinds. By watermarking, or laminating or embedding decorative elements in a handmade paper, you can control the amount and quality of transmitted light. Those areas with fewer layers will allow more light to pass through. Variegated textures and layers of color will enhance this effect.

Annette Sauermann: Installation Lichtfallen. Paper on steel string construction. 20ft. (6m), (above).

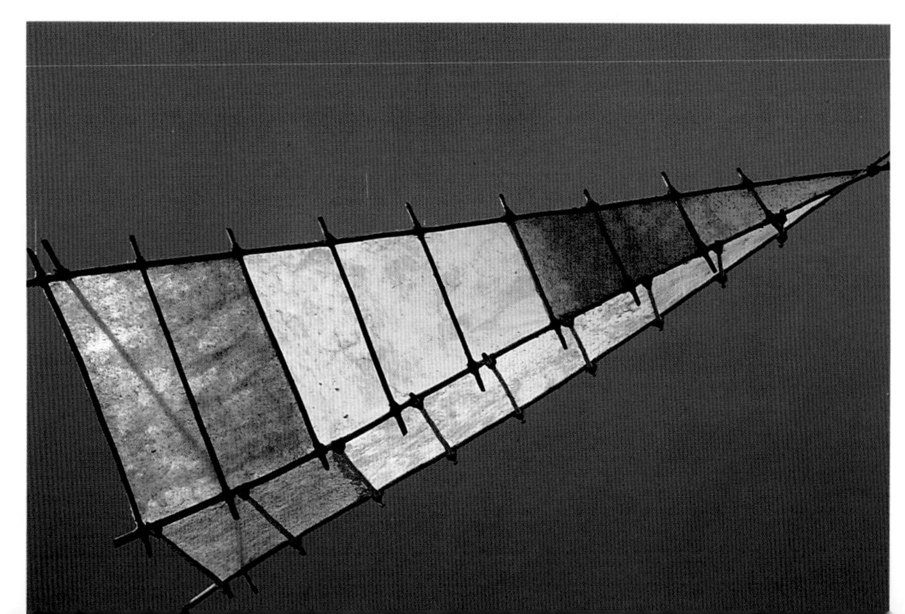

Jane Balsgaard: Paper Sculpture. Handmade paper and willow. 23½×39×17½in. (60×100×45cm).

PROJECT

Making a Candleshade

This project is based on an early European design. The shape is a simple one and can be copied by folding a sheet of plain paper around a small lampshade and cutting out a template. Scale the template down, if necessary, to fit the holder that will support the candleshade. Apply a fire retardant in the final stages of making the shade. Fire retardant aerosol sprays are available from most theatrical suppliers. They are manufactured for specific fabrics; be sure you use one which is suitable for paper. Follow the manufacturer's instructions for use.

- *prepared pulp: 2 colors*
- *mold and shaped deckle*
- *adhesive labels or similar (for experimental watermarks)*
- *couching cloths*
- *pressing boards*
- *adhesive: PVA or stick-flat glue*
- *fire retardant*

TECHNIQUES

- *sheetforming*
- *shaped paper*
- *experimental watermarks*
- *couching*
- *laminating*
- *pressing*

1 Make two deckles shaped according to your template. One should be slightly smaller along the curved outer edge. Make one sheet of paper using the larger deckle, and couch it onto a dampened felt. Then attach a watermark device to the surface of your mold within the smaller deckle shape.

2 Make another sheet with the smaller deckle, using a different color pulp.

3 Position the sheet on top of the larger piece, and couch it (see p. 57 for register).

4 **The longer curved edge of the blue piece sits slightly inside the larger gold sheet creating a simple eye-catching border.**

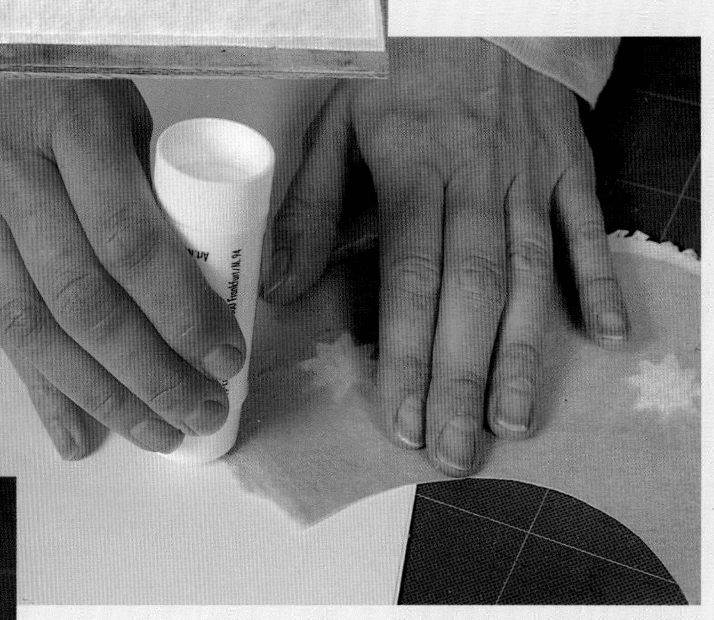

5 **When the pressed paper is dry, glue and join the edges.**

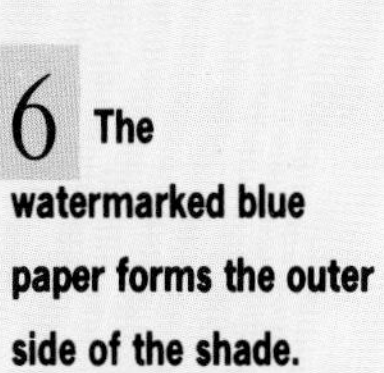

6 **The watermarked blue paper forms the outer side of the shade. Spray the finished shade with fire retardant and let it dry before lighting the candle.**

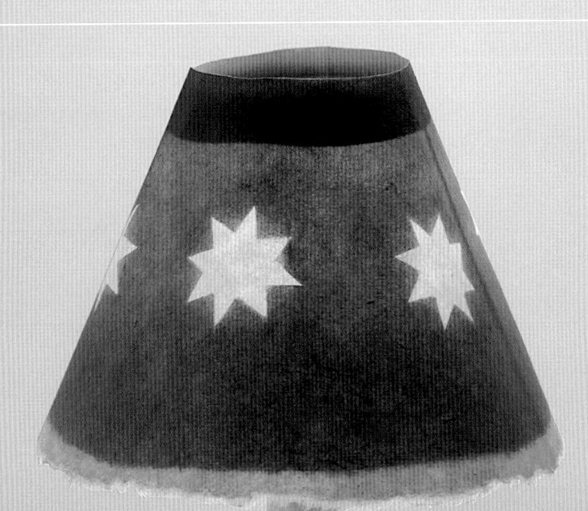

7 **When the candle is lit, the gold paper will shine through the design of stars.**

Karen Stahlecker: In Memorium: The Sacred Grove. Handmade kozo paper and mixed medium. 9×14ft. (2.7×4.2m).

Paper and Nature

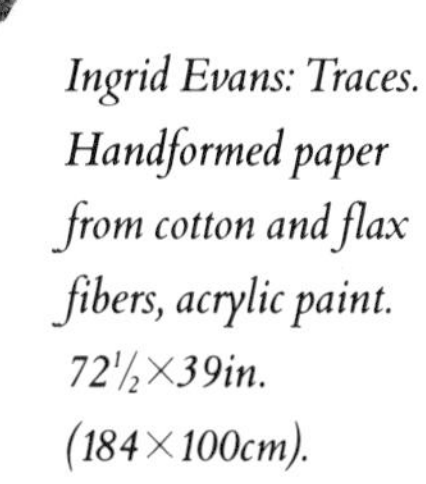

Ingrid Evans: Traces. Handformed paper from cotton and flax fibers, acrylic paint. 72½×39in. (184×100cm).

For many artists who work with paper, the transformative nature of the material and the cyclical process of papermaking are a direct source of inspiration. Papermaking itself, from the preliminary identification, gathering, and preparation of plant fibers to the creatively charged moment when a sheet is lifted from the vat, is perceived as a metaphor for life-growth cycles found in nature. The layering of sheets during couching and the simultaneous bonding of various elements during pressing are a reflection of geological events. The contradictory combinations found in nature are echoed in the fragility and strength, the smoothness and roughness, the translucency and opacity, the flexibility and structure of paper.

While the ritual traditions associated with paper and the meditative process of papermaking have appealed to many artists, paper's role as a natural reusable material has made it an attractive medium for addressing an increasing concern with more responsible use of the earth's resources. Throughout its history, paper has been a recycling medium, from the sheets made of hemp remnants and fishing nets in China to the use of ▶

Carol Farrow: To D.B. This Story Has No End. China clay book fired at 2372°F (1300°C). Glazed, rakud, sawdust fired slip cast base. 18×18in. (46×46cm).

◀ linen and cotton rags in Western papermaking. Today, old, discarded papers are digested into new creative forms and placed in fresh situations, transcending, yet never quite losing, the traces of a former existence.

CLAY PAPERS

There can sometimes be surprising affinities between natural materials – even if one is a mineral and the other a vegetable substance. For example, the translucency of porcelain clay and the rate of shrinkage as it dries can be compared to the same attributes in various plant fibers used for papermaking.

Paper and clay may seem unlikely companions, but they have a functional as well as an esthetic connection. Sometimes, China clay is used to coat the surface of a sheet of paper to make it more receptive to a particular printing method. Or it can be mixed with the pulp as a filler to make the paper more

PROJECT

Making a clay sheet

YOU WILL NEED

- *undyed pulp (partially prepared or recycled)*
- *finely ground clay (from ceramic suppliers) or local earth pigments*
- *mold and deckle*
- *scales*

TECHNIQUES

- *sheetforming*

2

1 **Weigh the clay before adding it to fresh water. You can adjust the percentage of clay to pulp according to weight. An approximate guide is 5 tablespoons of clay to 1lb. (454g) [dry weight] of fiber. This will vary, depending on the nature of the fiber.**

1

2 **If the clay is in solid form, break it into smaller pieces. Stir the solution occasionally until the pieces have dissolved.**

Anne Vilsboll: from the series: Tropism – Invisible Growth. Handmade paper in natural colors. 41×33½in. (105×85cm), (left).

Jane Balsgaard: Man. Paper stretched over steel rods. 118×79×59in. (300×200×150cm), (right).

opaque. Japanese clay papers – such as najio maniai-shi (gampi paper with clay) – are famous for their colors: green (kabuta-tsuchi), yellow (tamago-tsuchi), creamy white (tokubo-tsuchi), and brown and dark gray (jakame-tsuchi). "Tsuchi" means earth, and the colors come from mixing different local clays in the papermaking process. The sheets have a smooth, velvety surface which is characteristic of clay papers. ▶

3 **Strain off the liquid, leaving the residue of softened clay. Repeat this rinsing process several times, and then add enough fresh water to turn the clay into a thin slurry.**

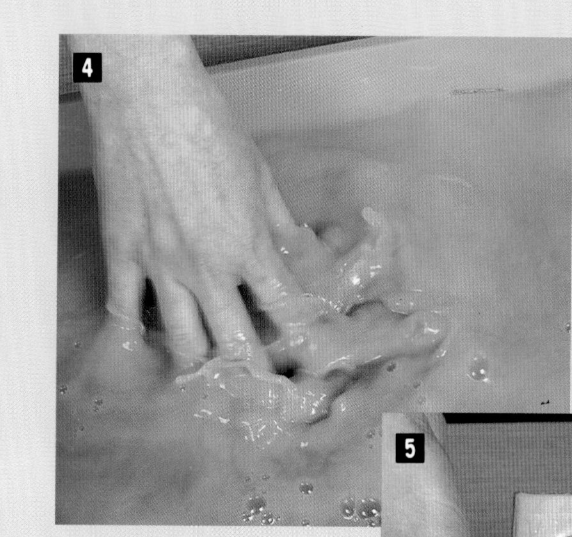

4 **Pour the liquid into the vat with the prepared fibers to make the pulp solution. Clay tends to settle at the bottom of the vat, so thorough mixing is essential.**

5 **Make your sheet of clay paper with the mold and deckle in the usual way.**

◀ VARYING TEXTURES

To make your own clay papers, you can experiment with different types of clay and, by varying the type and percentage of clay in the pulp, produce a range of surface textures from soft to hard, or fine to rough. For example, less finely ground clays will give a gritty surface to the paper. The nature of the fiber will also vary the character of the sheet.

If you are interested in the effects of natural pigments, you may wish to

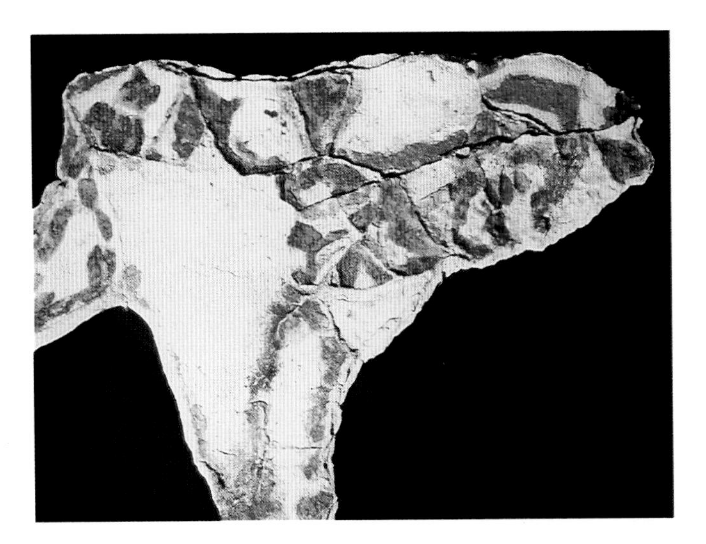

Golda Lewis: Irradiated Fragment (Great Jones St. series). Clay and paper. 19×16in. (48×40cm), (left).

Golda Lewis: Gift of the Spirit (East 16th St. series). Terracotta and paper. 22×30in. (56×40cm), (right).

PROJECT

Layering

YOU WILL NEED

- *colored pulps using natural earth pigment*
- *bandsaw*
- *sandpaper*
- *mold and deep deckle*
- *shallow tray*

TECHNIQUES

- *coloring and other additives*
- *pulp pouring*

1 **Cover a couching board with an absorbent cloth, and place it on a raised platform in a shallow tray. Position a mold and deckle, or a deeper frame (you can place several deckles on top of one another), in the middle of the board.**

2 **Strain each pulp into a container, and pour the first type into the frame, filling it to a height of about half an inch (12mm).**

3 **Let the water drain through the mold, and then use a sponge to remove the excess water and press the pulp into place.**

explore the tones from a range of clays or earth types. Each will have its own personality and history to contribute to the papermaking process, just as recycled papers and papermaking fibers do. Several types of clay are readily available from ceramic suppliers. Kaolin (China clay) is almost pure white, ball clays can vary from cream to gray, feldspars from pale pink to pale green. You might try clays which are prepared specifically for use as pottery glazes as slip deposits on the surface of your paper, or if your local soil has a high clay content, you could dig your own supplies.

Sometimes an undesirable element – for example, a high percentage of iron oxide – mitigates against the successful use of a certain clay. Darker clays tend to contain decayed vegetation which is likely to affect the quality and longevity of your paper. You can check the composition of a commercial clay with the supplier, or get a laboratory test done.

5 **When you have reached the top of the frame, remove it carefully and leave the work to air-dry in a cool place.**

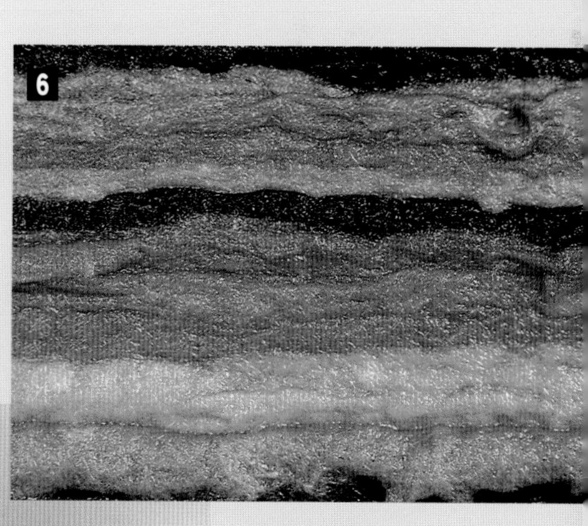

4 **Continue layering different pulps to build up successive levels within the frame. If you sponge out more water from some areas, you will create an undulating effect.**

5

6 **You can use a ceramic tool called a slab cutter to cut through the layers while the pulp is still wet or, once the piece has completely dried, use a bandsaw and fine sandpaper to sand the layers, bringing out the colors and cross-section effect still further.**

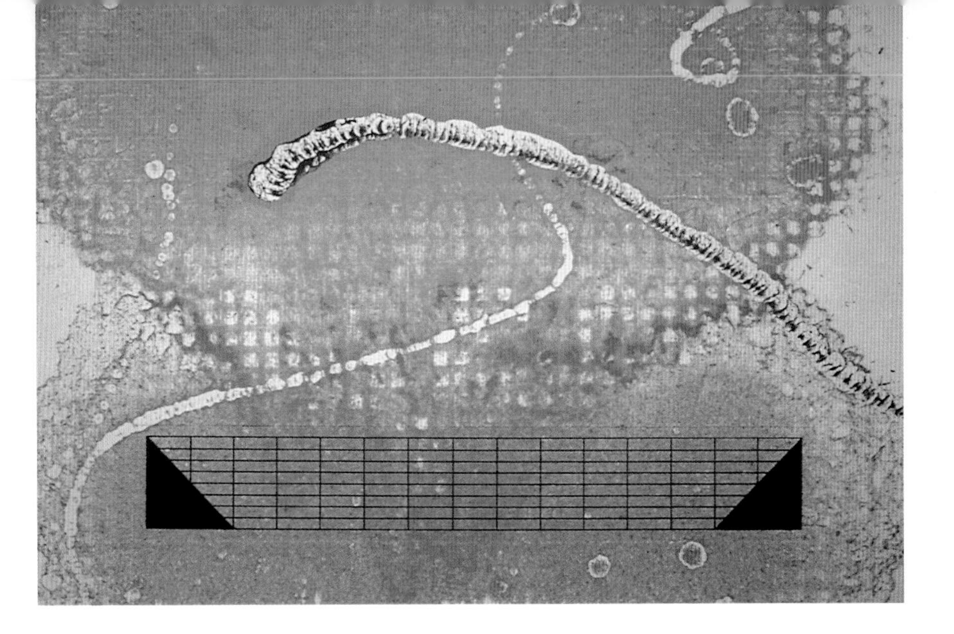

Laurence Barker: Untitled. Pulp painting with etching. 23×31in. (58×79cm).

Paper and Prints

Richard Royce: Crystal Traces. Cast paper with woodcut. 8×8×8ft. (2.4×2.4×2.4m).

Paper is a prerequisite for printing, and, as paper was invented in China, it is hardly surprising that the earliest examples of printing are Chinese. How and when printing was invented is uncertain. The use of carved personal seals, or "chops," is a very old Chinese practice. Small stamped images and ink rubbings taken from stone inscriptions (made possible by the earlier introduction of paper) during the Han dynasty (206 B.C.-A.D. 220) may have provided the impetus for the development of printing and with it the duplication of official texts and devotional images.

Many of the earliest surviving Chinese prints and manuscripts are religious. Printed prayer sheets bearing texts and depictions of Buddhist deities were produced from single, evenly textured carved woodblocks. The uncut surface of the block was covered with ink and the image or text transferred onto a sheet of paper. The print thus obtained – where the image is printed from the surface of the block and the background is cut away – is a relief print. It is one of four basic methods of printmaking, along with lithography, intaglio, and screenprinting.

The essence of printing is duplication, and printmaking techniques are mostly employed to make a number of identical copies from a single block or plate. A "monotype," however, is a unique print which cannot be reproduced, and the term is applied to any print process which does not allow for duplication.

The development of rigid surfaces other than wood or metal – masonite and acrylics, for example – offers a variety of experimental printing matrices for the modern printmaker to work with, and it is, of course, possible to print ▶

PROJECT

Monotype print

1 Here, a thin sheet of plexiglass has been chosen as both a drawing and drying surface.

2 You can trace a drawing onto the plexiglass before the transfer to paper.

3 Or, you can draw directly onto a Formica surface with water-based crayons and colored pencils, and use a dampened brush to create washes of color. Sumi inks, dry pigments, metallic powders, and watercolors will all create subtle marks and modulations of tone.

YOU WILL NEED

- *selection of handmade papers (as recommended)*
- *Formica-covered board or plexiglass (acrylic)*
- *water-based crayons*
- *colored pencils (Caran d'Ache)*
- *sumi ink (optional)*
- *aqueous dispersed and dry pigments (optional)*
- *couching felt*
- *rolling pin*

4 Try using plain sheets of paper made from a low-shrinkage pulp to begin with. A light sizing will prevent the water-based media from bleeding too far into the paper. The amount of water in the sheet and the degree of wetness of the plate surface will also affect the clarity and transfer of the image. Take a pressed, freshly made sheet and place it carefully over the monotype.

◀ an image without the use of a traditional printing press.

MATCHING PAPER AND PRINT

The choice of paper and the success of any drawing or print process are also clearly related. Artists have often looked for specific papers for specific prints: the paper often becomes an integral part of the image rather than an anonymous supporting surface. Handmade sheets offer a tremendous range of possibilities in the monotype process, where uniformity is not an issue.

The kind of fiber, the weight of the paper, whether or not the sheet has been pressed, how damp it is, and the type and amount of size used are all contributing factors. Using experimental watermarks, laminating translucent or colored shaped pieces of paper to a base sheet, or pulp painting onto a sheet before it is applied to a drawing will enhance the final appearance of the print.

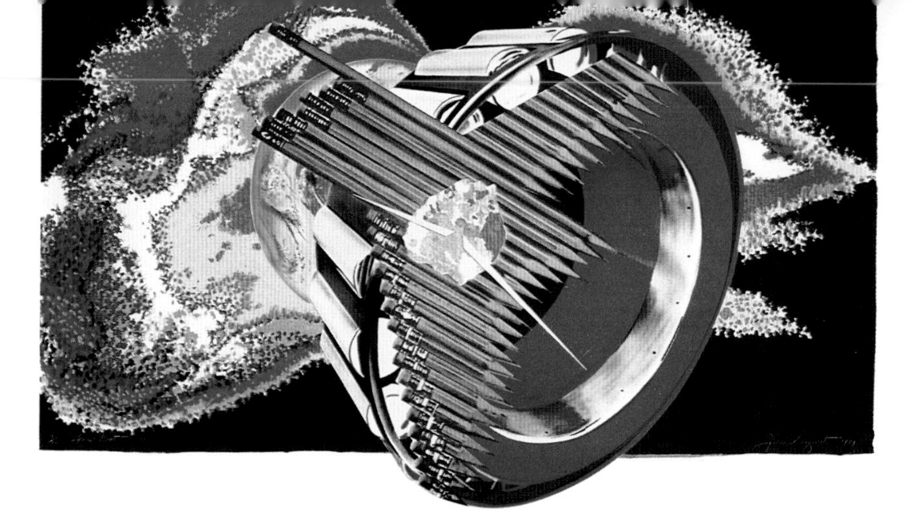

James Rosenquist: Space Dust. Colored, pressed paper pulp, collage elements, offset lithography. $62\frac{1}{2} \times 104\frac{1}{2}$in. ($158 \times 265$cm).

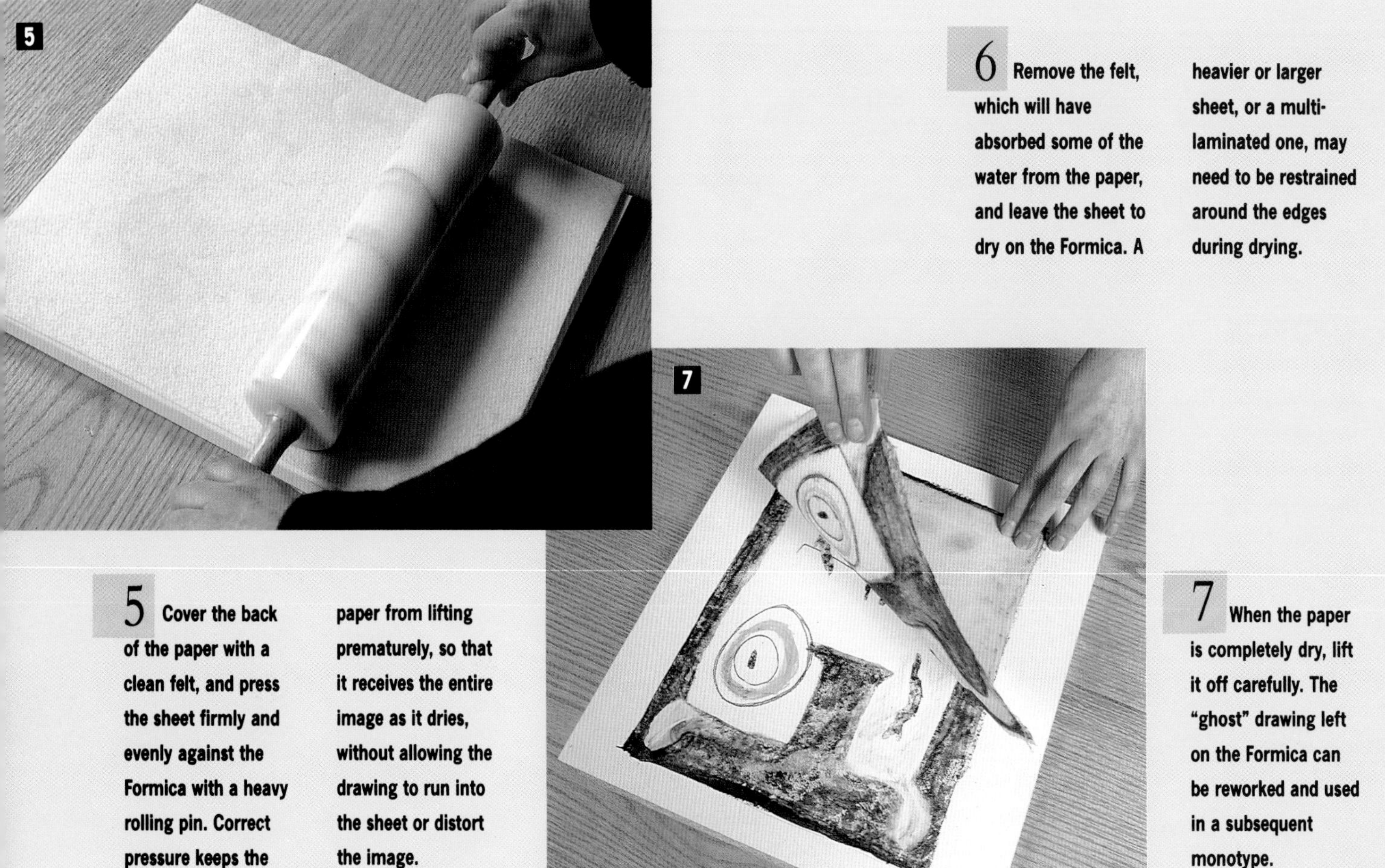

5 **Cover the back of the paper with a clean felt, and press the sheet firmly and evenly against the Formica with a heavy rolling pin. Correct pressure keeps the paper from lifting prematurely, so that it receives the entire image as it dries, without allowing the drawing to run into the sheet or distort the image.**

6 **Remove the felt, which will have absorbed some of the water from the paper, and leave the sheet to dry on the Formica. A heavier or larger sheet, or a multi-laminated one, may need to be restrained around the edges during drying.**

7 **When the paper is completely dry, lift it off carefully. The "ghost" drawing left on the Formica can be reworked and used in a subsequent monotype.**

Gallery

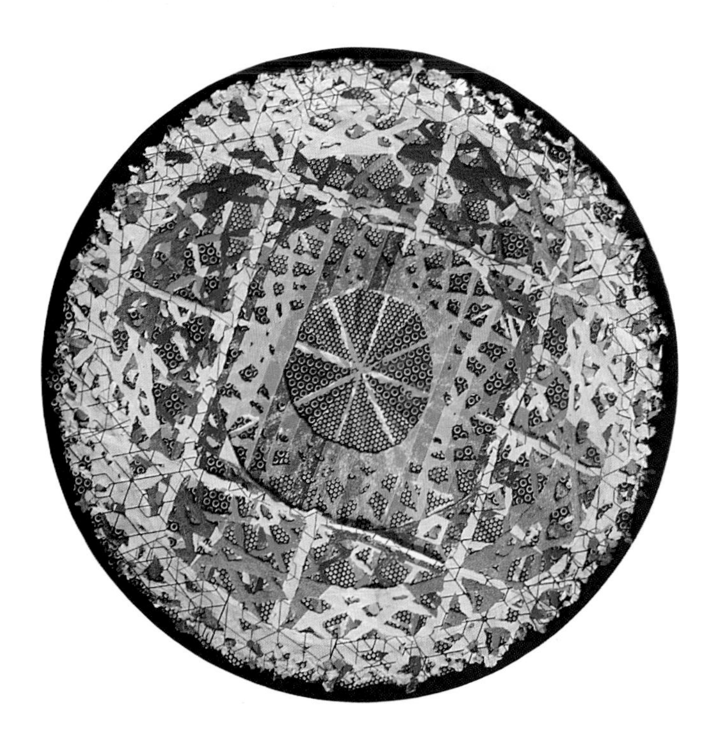

Alan Shields: Polar Route. Relief, screenprint, woodcut. 47in. (119cm) diam.

Peter Gentenaar: Blue Note. Etching (without press) on linen paper. 20×25½in. (50×65cm).

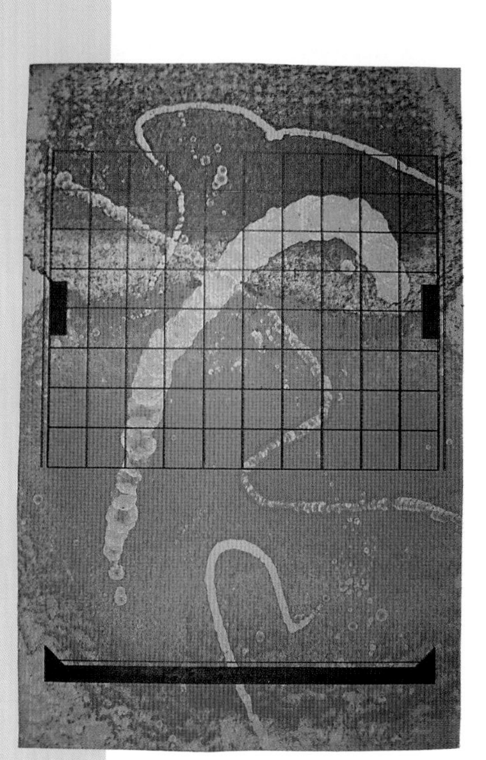

Laurence Barker: Black Diode. Etching on handmade paper. 40½×27in. (103×68cm).

Helmut Frerick: River-Action Paper (in memento mori). A symbolic enactment of the transitory nature of paper as a handmade paper triangle, 29ft. (9m) in size, is sacrificed to a torrential river.

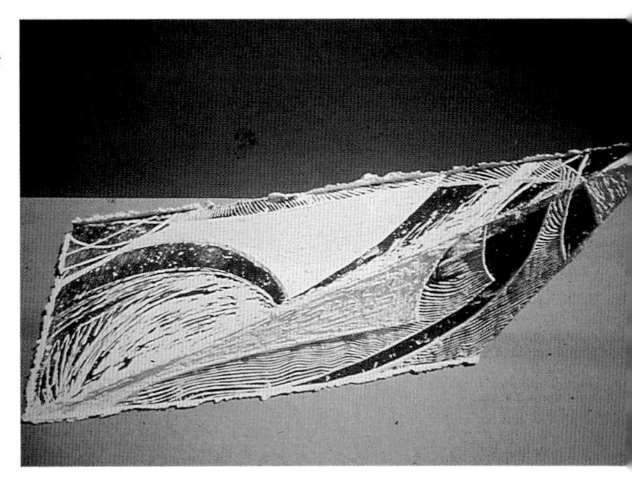

Richard Royce: Fragments of a Lost Civilization. Cast paper woodcut. 4×2×1½ft. (1.2×0.6×0.4m).

Anne Vilsboll: Domino. Handmade paper, plexiglass.

Puck Bramlage: Secret Life. Amate paper, Japanese silk paper, twigs, feathers

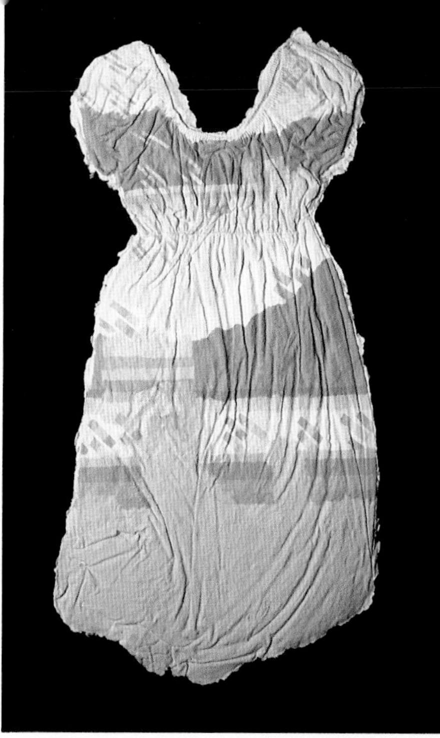

Margaret Ahrens Sahlstrand: Jans' Dress. Cast paper, pulp painting.

Ruth Millar: Book Nest. Handmade paper. 6×5×3in. (15×13×8cm).

GLOSSARY

ABACA Also called Manila hemp, a plant (*Musa textilis*) related to the banana and primarily cultivated in the Philippines. The leaf stems provide exceptionally strong fibers that make a versatile all-purpose papermaking pulp.

ALKALI A caustic substance used when COOKING plant fibers to remove gums, waxes, starch and other non-cellulose materials.

ALUM A complex salt (most commonly aluminum sulfate), used in the sizing of paper. Usually makes paper acidic and is therefore best avoided.

AMATE (AMATL) A pounded bark paper, similar to TAPA, originally using the inner bark of fig trees.

ASP (HORN) A notched piece of wood fitted into a platform across the vat against which the mold is placed for a moment to drain before couching.

BAGASSE Grass fiber from the stem of the sugarcane (*Saccharum officinarum*).

BAMBOO Grass fiber from the stem of the bamboo plant (*Phyllostachys aurea*), used for papermaking by the Chinese.

BAST Fiber derived from the inner bark of numerous shrubs and trees, including GAMPI and KOZO. The term also refers to herbaceous basts, such as FLAX and HEMP.

BEATING Separating and macerating fibers into pulp for sheet formation, done by hand or mechanical beater. See HOLLANDER.

BLEACHING A process used to purify and whiten pulp, usually done with chlorine compounds.

BLEED The spread or feathering of ink or color within a sheet of paper.

BONDING The capacity of cellulose fibers to adhere and combine. FIBRILLATION, HYDRATION, BEATING and drying promote bonding.

BUFFER (alkaline reserve) An alkaline substance (usually CALCIUM CARBONATE or magnesium carbonate) that helps protect paper from acidity in the environment.

CALCIUM CARBONATE Used primarily to promote longevity in paper (see BUFFER). In larger amounts it acts as a filler to retard shrinkage in paper casting, and in sheetforming to improve opacity and whiteness.

CALENDERING A pressing process used to impart a smooth or glazed finish to a sheet of paper.

CELLULOSE The chief component of plant tissue, which provides the basic substance for paper manufacture in the form of FIBER.

CHAIN LINES The widely spaced lines created by the retaining wires of a LAID MOLD and visible in the resulting paper.

CHINA CLAY (kaolin) A fine white powder that can be added to pulp as a filler and gives a smooth surface to the paper.

CMC See METHYLCELLULOSE.

COCKLING Surface undulation or wavy edges, usually due to uneven drying.

CODEX Manuscript volume, especially of a religious or classical text.

COLLAGE The pasting together of various materials to create an image.

COLLOGRAPH A relief or INTAGLIO process using a plate built up by collage.

COOKING The treatment of raw fibers to promote separation, remove contaminants, and dissolve unwanted plant material, usually achieved by heating in an alkaline solution.

COTTON The soft white filaments attached to the seeds of the cotton plant (Gossypium) that are one of the main fibers in Western hand papermaking.

COTTON LINTERS The pulp produced from the shorter seed hairs of the cotton plant which has been cooked, bleached, beaten and made into compressed sheets.

COTTON RAG A long-fibered pulp made from new rag cuttings.

COUCHING Transferring a freshly made sheet of paper from the MOLD surface onto a dampened FELT (Western papermaking) or directly onto the previously couched sheet (Japanese papermaking).

CUTTING An optional function of beating used to shorten the fibers.

DECKLE The removable frame which fits on and around the MOLD cover to contain the pulp and determine the size of the sheet.

DECKLE EDGE The distinctive, slightly ragged edge of a sheet of handmade paper, created by a small amount of pulp seeping under the DECKLE during formation.

DYES Soluble coloring agents which penetrate the structure of a fiber and become attached to it.

EMBEDDING Incorporating materials in a sheet of paper, so that the fibers hold the embedded material in place.

EMBOSSING The creation of a raised or depressed surface design in a sheet of paper.

FELT The woven woolen blanket onto which a newly formed sheet of paper is transferred, or couched, in traditional Western hand papermaking.

FERMENTATION Traditional method of loosening and softening cloth fibers prior to BEATING by leaving piles of wet rags to heat up and begin to rot. Also, the preliminary separation of plant fibers by soaking in alkaline solution or water, known as retting.

FIBER The slender, threadlike structure in plant tissue from which papermaking pulp is made.

FIBRILLATION The action of breaking up the surface of individual cellulose fibers during beating.

FILIGRANE (filigree) Ornamental work of fine wire formed into delicate tracery. See WATERMARK.

FILLER A general term for materials added to the pulp during beating to give the finished sheet a smoother, whiter or more opaque appearance by occupying some of the spaces between the fibers.

FINISH The surface qualities of a sheet of paper.

FLAX A bast fiber from the plant *Linum usitatissimum,* from which linen cloth is made. Pulp from flax fiber or linen rags makes exceptionally strong, translucent paper.

FLOATING MOLD In traditional Nepalese papermaking, a mold which rests on the surface of a vat of water.

FOREDGE The outer edge of a book, opposite the spine.

FORMATION AID A substance used to separate long-fibered pulps and control drainage during sheetforming. See NERI.

FURNISH The various ingredients used in making paper, particularly the fiber or blend of fibers used.

GAMPI A bast fiber from the plant *Wikstroemia diplomopha,* used for Japanese handmade paper.

GELATIN A type of SIZE made from animal tissue or bone.

GLAZING The gloss or polish of a sheet of paper and the process by which it is applied. See HOT-PRESSED.

GRAIN The orientation of fibers in a sheet of paper.

GUTTER In a book, the inner margins, or the center fold where two pages meet.

HALF-STUFF Any partially broken or beaten fiber. Further beating is necessary before it can be made into paper.

HEMP A bast fiber plant (*Cannabis sativa*) of high cellulose content, with a long history in Eastern and Western papermaking.

HOLLANDER A machine, invented in Holland in the late 17th century, for the preparation of rags or fibers for papermaking.

HOT-PRESSED (HP) Paper which is glazed by being pressed between hot, polished metal plates or heated metal rollers.

HYDRATION The absorption of water by the bruised fibers during beating.

IKAT Japanese ikat weaving, also known as kasuri, is a technique where lengths of yarn are tied and dyed before weaving.

INTAGLIO A method of printmaking in which the image is cut below the surface of the plate.

JUTE A bast fiber from the Asian plant *Corchorus capsularis*, widely used for making sackcloth, bags and cordage.

KAMIKO Japanese non-woven paper cloth.

KAOLIN See CHINA CLAY.

KAPOK A hollow, seed-hair fiber from the silk-cotton tree (*Ceiba pentandra*), used to stuff pillows and in padded clothing.

KOZO Japan's most widely used bast fiber, from the paper mulberry tree (*Broussonetia papyrifera*).

LAID LINES The closely spaced lines seen in paper made on a LAID MOLD.

LAID MOLD A mold whose cover is made of closely spaced, parallel wires or bamboo strips, held in place by more widely spaced, perpendicular chain wires or threads.

LAMINATING Combining layers of paper by couching one newly formed sheet on top of another to create a single sheet.

LIGNIN A component of plants that rejects water and tends to decrease fiber-to-fiber bonding in paper. It must therefore be removed before papermaking begins.

LINO BLOCK A means of block printing using linoleum instead of wood. See WOODBLOCK.

LITHOGRAPHY A method of printmaking where an image is made on a flat stone or plate using a material that attracts ink, whereas surrounding areas are treated to repel ink.

LYE A strong alkaline solution used to cook fibers.

METHYLCELLULOSE A specially formulated powder that can be used as an adhesive, as a surface size, or to promote fiber-to-fiber bonding and strengthen paper castings.

MITSUMATA Fiber from the shrub *Edgeworthia chrysantha,* a major source for Japanese papermaking.

MONOTYPE (or monoprint) A print, or print process, which does not allow duplication.

MORDANT In dyeing, a chemical substance used to fix colors applied to fibers.

MOLD A rectangular wooden frame covered with a sieve-like laid or wove wire surface, used for sheetforming.

NAGASHI-ZUKI Japanese term for the hand papermaking process using lightly beaten bast fibers and NERI.

NERI A viscous FORMATION AID in Japanese papermaking, usually derived from the roots of the *tororo-aoi* plant, a member of the hibiscus family.

NOT (Not Hot-Pressed) Slightly rough, traditional paper finish, made by pressing damp paper against itself after the first wet press between felts. Also known as cold-pressed.

PACK A pile of damp sheets, separated from the felts after the first pressing. Also, a small stack of paper ready for glazing.

PAPYRUS A laminated writing surface made from the sliced inner pith of the papyrus plant.

PARCHMENT A writing surface prepared from the skins of animals, especially sheep and goats.

PELLON Couching fabric.

pH A term used to denote the degree of acidity or alkalinity of a substance.

PIGMENT Coloring matter in the form of insoluble, finely ground particles, which have no affinity to the material they are coloring and must be used with a RETENTION AGENT.

POST The pile of newly formed sheets alternated with couching felts, ready for pressing.

PULP The aqueous mixture of ground-up refined fibrous material, from which paper is made.

QUIRE A set of 25 sheets of paper (originally 24).

RAFFIA Leaf fiber from a Madagascar palm tree (*Raphia ruffia*), used for cloth, hats, and baskets.

RETENTION AGENT A substance formulated to aid the binding of pigments, dyes and other additives to the fiber.

RETTING See FERMENTATION.

RICE PAPER The paper-like material made by cutting and pressing the pith of the rice-paper plant (*Tetrapanax papyriferus*).

ROSIN A sizing agent derived from the distillation of turpentine or from the treated gum pine trees.

ROUGH Traditional paper surface imparted by the weave of the felts during the first wet press.

SCREENPRINTING Printmaking achieved by blocking out areas of a screen or mesh and forcing ink or paint through the open areas onto the chosen surface.

SHIFU Japanese woven paper cloth.

SIGNATURE In bookbinding, a section of pages folded from a single sheet of paper. Also, a letter or number on the first page of each folded section to indicate the order of binding.

SISAL A leaf fiber from the tropical plant *Agave sisalana*, commonly used for cordage.

SIZE A substance added during beating or after drying to make paper more water resistant. Originally a solution of GELATIN, gum, or starch, now various chemical agents.

SPINE The back edge of a book where the sections are joined together at the folds.

SPUR A group of sheets dried together rather than singly.

STAMPER Early device (based on the action of a pestle and mortar) for reducing papermaking materials to a pulp. Replaced by the HOLLANDER.

STUFF Pulp ready for making into paper.

SU See SUGETA.

SUGETA The Japanese papermaking mold comprising the *su* (removable, flexible screen) and *keta* (hinged wooden frame).

SUMINAGASHI A Japanese marbling technique.

TAMEZUKI The Japanese term for Western sheetforming.

TAPA Polynesian word for a cloth-like paper usually made from the inner bark of the paper mulberry.

TEMPLATE A pattern, usually of thin board or metal plate, from which a similar design can be made.

ULTRAFELT A strong, non-woven polyester felt. Not recommended as a couching felt.

VACUUM TABLE A system of forming paper using a vacuum process to compress the pulp, usually consisting of a perforated table connected to a vacuum chamber.

VAT Container for pulp in which sheets of paper are made.

VELLUM A fine grade of parchment prepared from skin of calf, lamb or kid.

WARP Threads stretched lengthwise on a loom. See WEFT.

WATERLEAF Unsized paper.

WASHI The Japanese term for handmade papers.

WATERMARK (wiremark) A translucent area in a sheet of paper, usually created by attaching a fine wire design to the mold surface.

WEFT Threads woven across the WARP.

WIREMARK See WATERMARK.

WOODBLOCK Used for woodcuts, where the background is cut away, leaving the linear elements in relief, and for wood engraving, where the lines are incised, leaving the background in relief.

WOVE MOLD Any mold with a woven mesh surface.

INDEX

References to captions and illustrations are in *italic*.

A

abaca (pulp) 23, 24, 86, 88
alkaline ratio 28
alkaline solution, cooking fibers in 25, 26
alkylketene dimer 51, 52
alum 51
alum rosin size 53
amate paper 80-1
artists, and paper 15
Arts and Crafts Movement 120
Aztec paper 80

B

back mark 48
Balsgaard, Jane *126, 131*
bamboo, fermentation of *27*
bamboo paper, drying *48*
Barker, Laurence *95, 134, 137*
bast fibers 23, 27, *63*, 100, 102
  Japanese *24, 25*, 31
beating, by hand 31
Becker, Helmut *111*
beetroot paper *97*
Bell, Lillian A. *110*
binders 35
birch bark 9
bleaching 29
bleedthrough 52, 117
blenders, in hand papermaking 32, *33*
blotters, acid-free 49
books
  making your own 120-1
  as objects 118-19
box molds *101*, 106
Bramlage, Puck *138*
bridge *36*, 36
brighteners 35
Buttercut 69
butterfly books 114, 116

C

calcium carbonate 33, 35
calender rollers 14
Capone, Vito *55, 113*
carrot paper *80*
cast paper 100, *102-7*, 102-7, *110-11*
caustic soda 28
cellulose 18, 23
chemical (hydrogen) bonding 30, 102-3
chemicals, for cooking 28
chlorine bleach 29
chops (personal seals) 134
Clark, Kathryn *56*
clay envelopes 9, 114
clay papers 130-3
clay tablets 8, 114
clays, for pottery glazes 134
collage, and combining techniques 74, *75-7*, 76-7
colored papers *34, 35*
combined compositions 77
concertina folding 114, 116
cooking, of fibers 26, 27, 28-9
Coop, Gerry *74, 76*
cotton fibers *23, 24*
cotton linters 24, 106
cotton pulp 88
cotton rags *23*
coucher *38*, 38
couching *38*, 38-9, 41, *42-3*, 56, 94, 104, 129
  multiple *see* lamination
Crump, Kathy *85, 117*

D

de-watering *see* pressing
deckle *see* mold and deckle
decorative effects 65, *66, 67*
dried flowers 62
drying box 49-50
drying lofts 48, *49*
dyes, for coloring paper 35

E

embedding *16-17*, 59, *61*, 107
embossing 72, *73*
etching *137*
Europe, paper in 11-13
Evans, Ingrid, *Traces 129*
exchange drying method 44, *49*
extractives 26

F

Faerber, Judith *95*
Farrow, Carol *59, 129*
felts 38, 40-1, 47, 49, *73*
fermentation 25-6
fibers
  beating and pulp preparation 30-5
  for cast paper pulp 100, 102
  over-beaten 86
  preparation of 25-9
  sources of 23-4
  storage of 29
fibrillation 30
fillers 35, 130
finishing 13, 51, 52
Flavin, Richard *74, 76*
flax fibers *24*
flax pulp 86
floating mold effects 67
floating mold method 18, *36, 65*
flower press 64
flowers and foliage 62, *63-4*, 64
foamboard 106
formation aid 27, 39, 86, 88
Foudrinier machines 14
Frerick, Helmut *138*
fresh flowers 64

G

gelatin 51-2
gelatin sizing *52-3*, 53
Gentenaar, Peter *137*
Gentenaar-Torley, Pat *94*
Gerard, John *74, 76*
glazing hammer *51*
grain direction, paper 120-1
grass fibers 23, 27
Great Harris Papyrus manuscript 78
gutter fold *116*

H

"half-stuff" 24
Hall, Joan *69*
hand papermaking 8, 15, 18-19
  Japanese papers 56, *65*, 65
  Nepalese *65*, 67
  see also nagashi-zuki sheetforming; *washi*
hand polishing *51*
hemicellulose 31
hemp 13
Hockney, David *15*
hog 37
Hollander beater *14*, 14, *30*, 30-1, *31*
horn 36
"hot-pressed" finish *51*, 52
Houdin, Guy *111-12*
Howard Clark press *45*
huun 80
hydration 30
hydraulic jack *45*, 47
hydrogen peroxide *29*, 29

I

Ibe, Kyoto *96*
ink rubbings *9*, 134
insect/fungal attacks 53
interfacing, use of 41, *64*
internal (beater/engine, stock) sizing 51, 52

J

Jaffe, Jeanne *100, 111*
Japan, papermaking in 13, 56, *65*, 65, 126, 131
  see also nagashi-zuki sheetforming; pouch binding
Japanese books 116-18

K

kamiko robes 122
kaolin 35, 133
keta 20
kissing off *41*
kite paper *10*
knotter 37
Koretsky, Donna *88, 111*
kozo fiber, preparation of 25-6
kozo paper 122

L

lace paper *65, 67*
laid molds and paper 20
laminate casting 104-5
lamination 56, *57-8*, 136
layer 44
leaf fibers *23*, 23, 27
leaf stem fibers 23
Lewis, Golda *85, 132*
literacy, and paper 12
local plant sources, collecting and cooking 26-8
loft drying 50, 52
Lutz, Winifred *103*
lye *see* caustic soda

M

Manheim, Julia *7, 15*
marbling 65, *69*
Massaguer, Soledad Vidad *69*
maze 36
Mészáros, Géza *72, 85*
methylcellulose 27, 52, 102, 106, 107
microwave, for flower drying *62*, 62
Millar, Ruth *139*
monotype 134, 136
mold and deckle *12*, 18, 20-1, *22*, 38, *40*, 69
molds 13-14, *20*
Moulin, Richard de Bas *62, 64*

N

nagashi-zuki sheetforming 13, *18, 20*, 39, 125
natural substances, coloring early papers 34
neri 39

Norris, Julie *72, 96*
"not" finish *51,* 52

O

O'Donnel, Hugh *54*
oiled paper *10-11,* 10
oriental papers, drying of 48-9

P

pack 44
palm leaves *9,* 9
paper
  and books 114, *115,* 116-21
  double-sided *57*
  drying methods 48-50
  dual role of 10
  early, coloring of 34-5
  extending sheet dimensions *58*
  invention of 9
  and light 126, *127-8*
  mass production of 8
  matching to process 136
  migration of 10-12
  and nature 129-33
  and prints 134, 136
  recycled 24, 104, 129
  shaped *58,* 69, *70-1*
  and textiles 122, *123-5,* 124-5
  unusual 81
  uses of 8, 10
  versatility of 12-13
paper discoloration 53
paper fabric, woven 122, 124-5
paper fans *12*
paper lanterns 126
paper manufacture, changes 13-14
paper mill interior *13*
paper precursors 8-9, 78, 80-1
paper pulp, recycled 24
paper sculpture 100
papermaking machines, dry end 48
"papiers à inclusions florales" *62*
papyrus *9,* 9, 78, *79-81*
parchment 9, *10-11,* 12
pigments 35, 132-3
pistolet 36
plant fibers 26, *31, 32,* 130
plaster molds *101-2*
  see also sheet casting
Polynesian paper 78, 80
posts 38, 40, 44
pouch binding 116-18, *118-21*
pressed flowers 62, 64
presses
  function of 44
  home-built *45,* 46-7
  wooden-screw *45*
pressing *46-7,* 49, 129
  of gelatin-soaked paper 52
  and parting *44,* 44-7
pressing board, felt-covered *53,* 53
pressing methods 45-6, *46*
printing 12
printmaking techniques 134
projects
  collage shapes *75-7*
  creating a watermark *83*
  experimental watermarks *84*
  floating mold paper *68*
  fresh flowers and foliage *62, 63-4*
  fun shapes *70-1*
  Japanese pouch binding *118-21*
  lace paper *67*
  making a candleshade *127-8*
  making a clay sheet *130-1*
  making a mold and deckle *22*
  making a Papyrus sheet *79-81*
  making a plaster mold *101-2*
  monotype print *135-6*
  shifu 123-5
  silk thread *66*
  single-section book *115-17*
  speckled paper *66*
  vacuum forming *92-3*
  volumetric forms *109*
pulp
  beaten, storage of 33-4
  use of natural colorants 34-5
  versatility of 15
pulp casting 105, 106-7
pulp painting 85-6, *87,* 136
pulp spraying 88, *89-90,* 90

Q

quires 38

R

rag papers, early, colors of 34
rags 13, *23,* 25, *26,* 30
rakasui paper, with embedding *16-17, 59*
Ramsay, Ted *100*
recycling 24, 104, 129
Reese-Heim, Prof. Dorothea *97*
Reeves, Dianne L. *118*
registration mark *57*
relief print 134
relief surfaces 106
restraint drying 49
retting *see* fermentation
rice paper 81
rice straw *25*
Rosenquist, James 136
rosin 51, 53
Roth, Otavio *78,* 81
"rough" finish *51,* 52
Royce, Richard *134, 138*
Ryan, Paul *95*

S

safety, in handpapermaking 29
Sahlstrand, Margaret Ahrens *72, 139*
Sakurai, Sadako *122*
sarcophagus, Rameses III *8*
Sauermann, Annette *126*
Schei, Cathrine *110*
screen surfaces 21, *22,* 90
seed fibers 23, 100
Sellergren, Helena *126*
sheet casting *103-5,* 104-5
sheetforming
  and couching 38-43
  Japanese *see nagashi-zuki* sheetforming
Shields, Alan *15, 137*
shifu fabrics 122, *123-5,* 124-5
shrinkage, fibers 100, 102, 103, 104
sizing 13
  and finishing 51-3
slab cutter *133*
soda ash 28
Sowiski, Peter *56, 96*
special papers *65*
spirit money, Chinese *8*
spurs 48
  immersed in gelatin 51-2
Stahlecker, Karen *129*
stampers 13, *30,* 30
Stamsta, Jean *97*
starch size 51, 52
stay *36*
stencils 65, 86
stitched binding 116
su 20, 21, 39
sugeta 20, 39
suminagashi (Japanese marbling) *69*
surface sizing 52
  sizes for 51, 53

T

Tabbert, Josephine *8, 110*
tame-kuzi papermaking 18
tapa cloth *78,* 78, 80
techniques
  colored strips *61*
  couching *42-3*
  different effects *57-8*
  embedding an image *61*
  embossed shapes *73*
  pulp painting *86, 87*
  revealing an image *60*
  sheet casting *103-5*
  sheetforming *40-1*
  two-dimensional spraying *89-90*
textiles, and paper 122, *123-5,* 124-5
textured fabrics 72
Thayer, Nancy *107*
titanium dioxide 35
Tomasso, Ray *98-9*
tracing paper, recycled 104

U

uchigumo (cloud paper) 126

V

vacuum forming 91, *92-3,* 94
vacuum table *91,* 91, 94
vatman *38,* 38
vats 19, 36-7, *37*
vellum 12
Vilsbøll, Anne *95, 131, 138*
volumetric forms 102-4, *109*

W

washi (Japanese handmade paper) 10, 126
wasps' nests 81
"waterleaf" paper 51, 53
watermarks *82,* 82, *83-4,* 136
Watt, Sally Anne *124*
Weber, Therese *56, 96*
Weizsacker, Andreas von *96*
wet-dry vacuum system *92,* 94
Williams, Jody *110*
wood blocks 72
woodblock printed paper 65
woodblock prints *10*
wove molds and paper 20-1, *21*

Y

yucca 27

Acknowledgments

The publisher and author would like to thank the many artists and craftspeople who have loaned transparencies of their work, as well as the photographers and agencies listed below. For the handmade papers shown on page 65 the author would like to thank John Gerard, Berlin, Germany (paper with embedded pieces of colored paper and rag fibers) and Carriage House Handmade Paper, Boston, USA (Blue Cascade, Sugarplum Sparkle, Violet Moonlight); page 72: Sea Penn Press and Papermill, USA (for Fern and Salmonberry).

The author would like personally to thank Jon Wyand (for staging of demonstration photography), Sally MacEachern (for invaluable assistance) and to express appreciation to those friends and family whose patience, support and good humor underpin the writing of this book.

The author and publisher have made every effort to identify the copyright owners of the pictures used in the book; they apologize for any omissions.

Bibliography

Barrett, Timothy, *Japanese Papermaking – Traditions, Tools and Techniques.* Weatherhill, New York/Tokyo, (1983).

Bell, Lillian, *Plant Fibres for Papermaking.* Liliacae Press.

Cunning, Sheril, *Handmade Paper: A Practical Guide to Oriental and Western Techniques.* Raven's Word Press.

Heller, Jules, *Papermaking,* Watson Guptill, New York, (1978).

Hughes, Suki, *Washi: The World of Japanese Paper.* Kodansha International, Tokyo, (1978).

Hunter, Dard, *Papermaking – The History and Technique of an Ancient Craft,* Dover Publications, Inc., New York.

Ikegami, Kojiro, *Japanese Bookbinding, Instructions from a Master Craftsman,* adapted by Barbara Stephan. Weatherhill, New York, (1986).

Johnson, Arthur W, *The Practical Guide to Craft Bookbinding,* Thames and Hudson, London, (1985).

Koretsky, Elaine, *Colour for the Hand Papermaker.* Carriage House Press, Brookline, MA, (1983).

Mason, John, *Paper Making as an Artistic Craft,* Faber, London, (1959). Revised edition: Twelve by Eight Press, Leicester, (1963).

Rudin, Bo, *Making Paper – A Look into the History of an Ancient Craft.* Rudins, Sweden.

Saddington, Marianne, *Making Your Own Paper – An Introduction to Creative Paper-making.* New Holland, London, (1991).

Stearns, Lynn, *Papermaking for Basketry.* Press de LaPlantz Inc., Bayside, Calif. (1988).

Studley, Vance, *The Art and Craft of Handmade Paper.* Studio Vista, London, (1978); Van Nostrand Reinhold Company Inc., New York, (1977).

Toale, Bernard, *The Art of Papermaking.* Davis Publications Inc., Worcester, MA, (1983).

Turner, Sylvie and Skjold, Birgit, *Handmade Paper Today.* Lund Humphries, London, (1983).

Credits

KEY: **T** – TOP **B** – BELOW **C** – CENTER **L** – LEFT **R** – RIGHT

2 Stuart Baynes/Wookey Hole Papermill; **6** The Mansell Collection; **7** Photo by Julia Manheim/ Keith Morris; **8TR** Sophie Dawson; **8TL** Josephine Tabbert; **8BR** ET Archive; **9** Puck Bramlage; **9BR** Sophie Dawson; **9BL** Sophie Dawson; **10T** Ino Paper Museum; **10B** Sophie Dawson; **11T** Sally Anne Watt; **11B** ET Archive; **12T** Sally Anne Watt; **12B** Sophie Dawson; **13** ET Archive; **14T** Sophie Dawson; **14B** The Mansell Collection; **15T** Alan Shields/Tyler Graphics Ltd; **15C** David Hockney/ Tyler Graphics Ltd; **15B** Photo by Julia Manheim/ Ed Barber; **17C** Sophie Dawson; **18T** Ino Paper Museum; **18B** Stuart Baynes/Wookey Hole Papermill; **23T** Sophie Dawson; **23B** Sophie Dawson; **24T** Sophie Dawson; **25T** Judith Sugarman; **25B** Ino Paper Museum; **26** Moulin â Papier, Richard de Bas; **27** Judith Sugarman; **30T** Moulin â Papier, Richard de Bas; **30B** Sophie Dawson; **31T** Sophie Dawson; **31B** Peter Gentenaar-Torley; **34** Sophie Dawson; **36T** Thomas Kelly/The Body Shop International; **36B** Sophie Dawson; **38T** Ino Paper Museum; **38B** Stuart Baynes/Wookey Hole Papermills; **39T** Kathryn Clark/Twinrocker; **39B** The Mansell Collection; **44T** Moulin â Papier, Richard de Bas; **44B** ET Archive; **45L** Sophie Dawson; **45BR** Lee S McDonald; **48T** Judith Sugarman; **48B** Sophie Dawson; **49** Sophie Dawson; **51T** Sophie Dawson; **54** Hugh O'Donell/Tyler Graphics Ltd; **55** Vito Capone; **56T** Kathryn Clark/Twinrocker; **56C** Peter Sowiski; **56B** Therese Weber; **59T** Sophie Dawson; **59B** Carol S Farrow; **62T** Moulin â Papier, Richard de Bas; **62B** Quarto Publishing plc; **64T** Moulin â Papier, Richard de Bas; **69TR** Soledad Vidal Massaguer; **69C** Sophie Dawson; **69BR** Joan Hall/photo by Hal Bundy, courtesy Elliot Smith Gallery; **72T** Meszaros Geza; **72C** Julie Norris; **72B** Margaret Ahrens Salstrand; **74T** Gerry Copp/J Copp; **74C** John Gerard; **74B** Richard Flavin; **76T** John Gerard; **76C** Gerry Copp/J Copp; **77T** Richard Flavin; **78T** Otavio Roth; **80** Papyrus Institute; **82T** The Bank of England; **85T** Golda Lewis; **85C** Meszaros Geza; **85B** Kathy Crump; **88T** Elaine Koketsky/Sophie Dawson; **88B** Donna Koretsky/photo by the artist; **91T** Sophie Dawson; **91B** Sophie Dawson; **94** Pat Gentenaar/Torley; **95T** P Ryan; **95C** Judith Faerber; **95BL** Anne Vilsbøll; **95BR** Laurence Barker; **96TL** Julie Norris; **96TR** Kyoko Ibe; **96CL** Andreas Von Weizsacer; **96CR** Therese Weber; **96B** Peter Sowiski/photo by the artist; **97T** Dorothea Reese-Heim; **97BL** Jean Stamsta; **97BR** Papyrus Institute; **98-99** Ray Tomasso – Inter Ocean Curiosity Studio; **100T** Jeanne Jaffe; **100B** Ted Ramsay, Huron River Studio USA/photo by the artist; **103T** Photo by Winifred Lutz and prop. of Winifred Lutz; **105T** Lilian A Bell/photo by David Browne; **107T** Nancy Thayer; **110TL** Jody Williams/photo by the artist; **110TR** Lillian A Bell/David Browne; **110CL** Cathrine Schei; **110BL** Josephine Tabbert; **110BR** Ted Ramsay, Huron River Studio, USA/photo by the artist; **111TL** Jeanne Jaffe; **111TR** Donna Koretsky/ photo by Dana Salvo; **111B** Helmut Becker; **112-113** Guy Houdouin, L'Imprimerie/Adam Rzepka; **114** Vito Capone; **114C** Sophie Dawson; **117** Kathy Crump; **118T** Dianne L Reeves; **119T** Dianne L Reeves; **122T** Sadako Sakurai; **122B** Sadako Sakurai; **124** Sally Anne Watt; **126T** Helena Sellergen; **126C** Annette Sauermann/Anne Gold; **126B** Jane Balsgaard; **129TL** Stahlecker; **129TR** Stahlecker; **129C** Ingrid Evans; **129B** Carol S Farrow; **130** Anne Vilsbøll; **131T** Jane Balsgaard; **132T** Golda Lewis; **133T** Golda Lewis; **134T** Laurence Baker; **134B** Richard Royce/Atelier Royce, Salem, Mass., USA; **135-136** Kate Robinson; **136T** James Rosenquist/ Tyler Graphics Ltd; **137T** Alan Shields/Tyler Graphics Ltd; **137BL** Laurence Barker; **137BR** Peter Gentenaar-Torley; **138TL** Helmut Frerick; **138TR** Richard Royce/Atelier Royce, Salem, Mass., USA; **138BL** Puck Bramlage; **138BR** Anne Vilsbøll; **139TL** Margaret Ahrens Sahlstrand; **139TR** Ruth Millar.

Useful addresses

Centro de Investigaciones
en Fibras Y Papel (CIFP)
(Centre for Fibres and
Paper Research)
c/o Lil Mena
Apartado 103-1002
San Jose, Costa Rica

The Hall of Awa Japanese
Handmade Paper
141 Kawahigashi
Yamakawa-cho, Oe-gun
Tokushima Prefecture
Japan 779-34

Hand Papermaking (Journal)
PO Box 77027
Washington, DC 20013-7027
USA

International Association
of Hand Papermakers & Paper
Artists (IAPMA)
Secretary: Pat Gentenaar-Torley
Sir W. Churchillaan 1009
2286 AD Rijswick
The Netherlands

Papermakers of Australia
Papermill
University of Tasmania
GPO Box 252C
Hobart, Tas. 7001
Australia

Papermakers of
South Africa
c/o John Roome &
Tony Starkey
Paperworks Unit
Department of Fine Art
Techknikon Natal
953 Durban 4000
South Africa

Currently there are no specialist papermaking retailers in the UK. Supplies can be obtained from most good general art suppliers. If you would like any further information please send sae to Sophie Dawson, Vanbrugh Castle, Maze Hill, Greenwich, London SE10 8XZ